Title Page

Bundal Log Identification Not In Use

Tubal's Recipe for Physics

Imagining Two Temporal Dimensions

D.R. Poeppelmeyer
Original Bundal Log Provided by Tubal

Available in both book and Kindle format

Copyright Page

Bundal Log Identification Not In Use

Dedication

Author's Dedication:

This translation might as well be a nyctograph. It is therefore dedicated to Amazon.com who allows so many of us authors to whistle in the dark.

Composers' Dedication:

"When human beings learn how to manipulate reality through temporal engineering, the real God is going to look like a saint."

Tubal's Dedication:

"We are beginning to believe in you. Eat up."

Table of Contents

Note: Unlike other works of 19th and 20th century metaphysics, great care was expended not to create neologisms. Nonetheless, we created an acronym that could be used to represent either a noun or a verb. The acronym is SHFTTS (Seeing, Hearing, Feeling, Tasting, Smelling and Sensing). This is explained in a chapter by the same title, SHFTTS.

Table of Figures

How Much Space Do We Need?

What if there was a simple idea that would unify Relativity and Quantum Mechanics? What if the idea were as simple as describing the color red to a blind person? The idea is simple. Describing the idea to a blind person would be difficult.

This book describes a bi-temporal world to mono-temporal humans. The simple idea is we live in a bi-temporal, bi-spatial universe. There are two spatial dimensions. There are two temporal dimensions.

Traditionally we have always thought of ourselves as inhabiting three spatial dimensions. It is difficult to conceive how we might actually inhabit only two spatial dimensions, until we can imagine that distance has more to do with time than with space.

For millennia we humans have imagined our three spatial dimensions existing within time. Then Einstein combined space and time into a unified Hyperspace. String physics currently requires numerous additional spatial dimensions in a frustrated effort to combine gravity into the three unified forces.

Scientific imagination has previously described life in two, three, and four dimensions. The popular work of Hinton and Abbott, beginning in the late 1880's, allowed readers to imagine a fourth spatial dimension. Such work paved the way for a general notion and acceptance of more than three dimensions.

This book does for time what Hinton and Abbott did for space. It attempts to imagine a second, temporal dimension. This is far more

1

difficult a task than simply imagining a tesseract. It is akin to explaining the color red to a blind person. The task is not impossible.

My own journey into imagining a second, temporal dimension is not described in this book. Rather, the results of my journey are described.

This book takes the form of an alien logbook. It pretends aliens have a bi-temporal physics that needs to be explained to us mono-temporal humans. A first read of this book may fill you with frustration and confusion as your operant paradigm collides with the aliens' in mutual misunderstanding. It is not always easy reading.

The best way to read this book is just to read it. There are whole parts of this book that may not be understood on a first reading. Our alien friends do not think linearly, since they are aware of a second temporal dimension. Translating their work into human experience can only result in what appears to be bits and pieces from our linear perspective. The composers of this book have worked hard to put the bits and pieces together into discernible music designed for the human ear.

The best I can describe this book is that it is for people who are conversant with German Enlightment philosophers. This audience includes physicists who are self-aware their assumptions have *apriori* components that guide their perceptions. This work also offers a strong critique for the relatively high evaluation mathematics plays in twentieth-century culture. The work is also useful to science fiction writers seeking a unique platform from which to view reality.

This work requires great effort to read and digest.

When you are done you will have successfully navigated an alien physics and you will be able to imagine a world with two temporal dimensions. Your bi-temporal imagination will take you to a world where only two spatial dimensions are necessary for understanding our existence.

Relax. Our world is much simpler than an infinite number of multiverses. In fact, it may be much simpler than three spatial dimensions.

Exotic Erotica

A burning, smelly, bundle of tubes passed by a crowd of other Bundals. Two larger Bundals fired a waxy, tarlike substance at the little Bundal.

> Smaller Bundals are seen as somewhat purer in regards to chemical and electrical manufacturing.

The smaller Bundal started to roll on the ground as if knocked over.

Numerous lights, smells, and sounds were emitted. All were faint and soft. One of the larger Bundals reached out and some of its tubes began to unwind, growing straighter.

A dance of insertion and reception began as the tubes began to intertwine among the tubes of the smaller Bundal. The third Bundal began a similar process as the smallest Bundal received its brightly colored tube.

It was a multiple hour dance but no human would be able to discern the number of Bundals intertwining.

The intertwined Bundals looked like several trees in a windstorm. They stood up and grew together. Some tubes seemed passive and relaxed. Other tubes seemed aggressive, wildly flailing for contact or release from contact.

Over the years a few other Bundals would join the formation.

The colors and smells and pulsations from the many tubes would grow strong and loud. The sounds of beating drums would emanate from the depths of the tangled, rhythmic mess. Fluids, gases, and smells would be injected into the surrounding environment covering the actors with mists of unknown compositions. The tubes would vibrate.

After a few hundred earth-years, the tubes would disengage. The process of disengagement was always rapid.

Sometimes tubes would tear and cries of repair would begin to form.

Make no mistake. The number of Bundals to separate was always greater the number of Bundals that came together.

It might take a few years for all of the Bundals to regain their equilibrium, or their own normal processes.

In the end the Bundals were always pleased. The living universe had coalesced to form another low entropy environment. Of-course, the mess left behind was high entropy.

New Bundals were born.

Warning: This chapter is not suitable for Bundals under three-hundred earth-years old.

Author's Guarantee on Op Narrative

I have a wonderful Indonesia recipe for chili. It involves goat meat and lemon grass stalks.

Yesterday, I had samples of thirteen dishes prepared by a Korean chef. Each was delicious. Each dish used seaweed as an ingredient. Unless you were born near the sea, you would think all seaweed tastes the same. The chef informed me she used eight types of seaweed, each with its own flavor and texture. Nothing tasted the same. Each was delicious.

Tubal's recipe is musical in nature. Translating Tubal's music into English narrative has proven difficult. Describing the use of exotic ingredients used in unconventional ways has proven more difficult. Attempting to translate physics into common language has added additional spice to this recipe.

The best that an author can do, given any confluence of words, is to guarantee a final product. Regardless of how you evaluate this recipe, ample provision is made for you to taste the existence of two temporal dimensions.

As a beginning chef you know the following.

You can imagine a point, a line, a plane, and a cube.

You can imagine a cube. It has three dimensions.

You can imagine a plane. It has two dimensions.

You can imagine a line. It has one dimension.

You can imagine a point floating somewhere in space.

Mathematicians tell us points have zero dimensions. Now how can you claim to imagine something with zero dimensions? It seems impossible, but you can fool yourself into thinking points have zero dimensions.

If you can imagine a point with zero dimensions, you can imagine a second temporal dimension. I guarantee it.

Please note this guarantee is not a warranty. There are all sorts of translation problems.

How do you edit a Jackson Pollock painting? Do you remove one of the black dribbles? Do you shorten a red stripe? Do you add a dash of orange to balance the commentary? Do you salivate over it?

Take something as simple as a piece of Shakespeare. Humans edit Shakespeare with each and every presentation. Each passing century dulls his comedic and witty elements as we are unacquainted with sixteenth century idiosyncrasies, transgender love notwithstanding. Our loss of the sixteenth century context removes us from the nuances necessary to laugh at all parts of his plays. Even today, producers, playwrights, and actors recite Shakespeare by removing and changing some words or lines. Such changes arise through forgetfulness, misprinting, or are made because they are extraneous to the needs of the audience.

When you hear Shakespeare, you are not hearing Shakespeare. You are hearing an interpretation of Shakespeare sanitized just for you.

Shakespeare is easy compared to Bundal music.

Bi-temporal thinking does not translate well to linear thinking. Few items will just flow from one chapter to another. Quantum mechanics demands the ideas to jump around.

There will be much preparation required before even starting to get into the ingredients of the recipe. Every kitchen needs put in order. The opening chapters are necessary in order to prepare the reader before indulging on the main course.

Tubal's Recipe for Physics is an alien composition written as op-narrative. It is based on a unique hermeneutic: The recipe combines substance with event to create phenomenon.

> Another way of saying this is the hermeneutic represents a synthesis of a substance-ontology with an event-ontology.

> The Bundals consider this the combining of nouns with verbs.

This recipe lets you put some of the ingredients together in different orders at different times. The composers believe the current order is best suited to human taste.

Read and reread any chapter you like at any time. How will you know you like it until after you have tasted enough?

This recipe involves exotic ingredients. Explanations defining the exotic materials are available at the library and on the internet. It will require a diligent amount of effort to actually acquire all of the ingredients used in this recipe.

> The truth is this recipe is vomited as index cards from within the Bundal log. The three-dimensional cards are musically arranged as a puzzle in two-dimensional, op-narrative form.

> You may cringe at the idea of tasting Bundal vomit.

> Will you actually eat spit? You do it every day.

> The best honey on earth is bee spit. Be assured. Alien vomit is a delicacy only the dedicated will savor.

You can still read the recipe without possession of all the ingredients. Only the after-taste may change.

If you acquire only some of the ingredients, the recipe is still edible without having everything. It just will not be as spicy.

Following the main course are chapters describing why the recipe should taste good to humans. It turns out our ancestors had a pretty good sense of taste. They knew their reality better than we know ours.

As you actually try to make this recipe, please know that different chefs may use the same recipe. Yet, each dish will taste differently depending on who makes it.

Granting the human race this delicacy is the purpose of this work.

D. R. Poeppelmeyer, Author

The Briggs-Meyer Inventory cites our Author as an INFJ.

Can You Find the Hermeneutic?

Reserve for Bundal Log Identification

The following visual is for people who already feel like they cannot understand this book. Can you find the hermeneutic? The hermeneutic may run horizontally, vertically, or diagonally.

Figure 1: Find the hermeneutic.

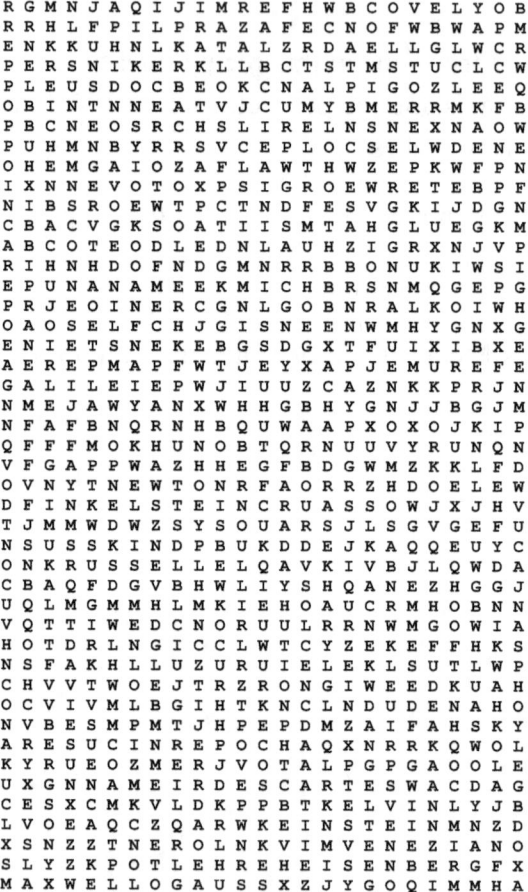

Tubal's Preface

For eons I doubted the existence of humans. All forms of life, as we Bundals know it, are directly evolved from the universe. Life knows life. Life knows its universe.

How can any creature not sense the entire universe? Such a creature can only be assured of itself.

Bundals evolved directly from light and gravity. We live in gravity and light. Space and time are derivatives. We make and use space and time much as you use fingers and toes. You may SHFTSS us in space and time but we do not look nor function as you expect. (SHFTSS is explained elsewhere.)

Having met humans we now know that life can evolve indirectly from the universe. Such life evolves only from space and time instead of gravity and light and is indirectly nurtured by gravity and light.

All Bundals know space and time have a limited existence.

Life forms evolving from space and time would be like the ghosts of the universe.

- Hardly here or there.
- Hardly this or that.
- Hardly determinate or indeterminate.
- Hardly black light or white light.
- Hardly past or future.

Space is not perpetually solid. Even humans know this.

Time is not eternally energetic. Even humans know this.

Such life would enjoy a brief existence of low entropy within a drop of gravity fluid known only to humans as space and time.

Why would we bother with such life forms and treat them on an equal basis with directly evolved life?

God does not use a stop-watch. What is the point of a brief existence?

I studied you to learn the anomaly of space and time creatures. You are rare indeed.

Imagine my surprise when I discovered that humans are not only relegated to space and time for SHFTSS. They cannot even SHFTSS all of space and time.

Bundal Joke:

Question: How does a human flower live in a desert without rainfall?

Answer: It pretends there is rain.

Humans make stuff up. What they do not know they pretend to know with stories of their own creation. Fictional stories are believed in every human discipline.

Every human being lives a myth, a combination of fiction and incomplete non-fiction.

Bundals would expect death to be the great non-fiction for humans. The opposite is true. Death seems to accelerate the telling of fictional stories about life.

Human cultures compete over which made-up stories are true. I pick one small story from all of your disciplines to SHFTSS why Bundals have difficulty in believing in your existence.

The Fictional Sale of Manhattan Island:

None of the following will make sense unless you know the history of the sale of Manhattan Island to the Dutch-West India Company. This story is an example of the many stories humans tell each other. The so-called facts in the story make little sense to Bundals. The following issues illustrate our confusion over typical human stories.

Bundals do not believe humans own land. (Do you?)

The Lenape Indians did not really believe the Dutch-West India Company could own Manhattan Island. They did not imagine they could sell God's land. (Do you?)

Bundals do not believe humans even knew their own world at the time of the story. (Do you?)

The name of the Dutch-West India Company ought to tell most humans someone got the continent wrong. Is Manhattan near India? (Are you listening?)

While Bundals can eat money, humans do not. (How valuable is Dutch money to Indians who live 3,600 miles from Holland?)

And what would the Indians do with sixty guilders anyway? (Melt them for arrowheads?)

So many parts of this story are farcical. Yet this fiction led to the establishment of a country that would rule world events throughout earth's twentieth-century. (Do you smell anything odd in the taste of this story?)

Of-course, the English at Jamestown tell a different story. The Virginia king was saved by the Algonquian queen and the rest is Pocahontas history.

The desert flower lives with pretend water.

Humans might laugh at the way Bundals misinterpret them. I will leave it to you to see how much Bundals misinterpret.

Twentieth-century geological explanations for the existence of Manhattan Island are equally enigmatic to Bundals. The Indian stories, whether Dutch or English, are as laughable as scientific geological stories.

Bundals quiver in confusion. Are humans going to claim entire galaxies as their own?

Humans could save time by simply planting a flag on earth and proclaiming: "We claim this universe for the sole possession of the human race."

> Humans now pay their political and corporate leaders to attach their personal names to stars and galaxies. Human stories of galactic conquest are as romantic as Pocahontas. Love and war make for kingdoms.
>
> Humans use to name the stars after their gods. They still do. Only now they use their own names.
>
> Bundal Joke:
>
> Question: Why do humans kill each other?
>
> Answer: Because some claim to own their part of the universe.

Humans should have exterminated themselves long ago.

- Human: The wall is not there. I choose to walk through it.
- Bundal: It will hurt.
- Human: We are the center of the universe.
- Bundal: It will hurt even more. Your entropy should become quite high long before your drop of space and time is reabsorbed in the gravitational fluid.

Bundals now believe in the existence of humans. Yet you remain a story difficult to believe.

We suspect your ignorance of time has enabled you to survive as long as you have. Your technological manipulation extends only to space. Space holds little danger compared to time.

We hope to give you the ability of learning how to do temporal engineering.

Using gravity and light your technologies will extend to temporal manipulation. You will have minimal effects on those of us who are directly evolved from the universe. You will have major effects on those who are evolved from space and time.

You will change yourselves as you change space and time.

Do not fear for Bundals. We are safe as creatures whose existence is directly nurtured by gravity and light.

What happens when a desert flower learns how to do temporal engineering instead of mere spatial engineering?

The only thing a spatial engineer can do is change a location, move things around in space. So a spatial desert flower can only replant itself in a different spot?

Big Woo Hoo song.

When the desert flower learns how to do temporal engineering, it can change space itself.

If the newly empowered desert flower wants lots and lots of water then look out Noah.

If the newly empowered desert flower is completely satisfied with pretend water and seeks to dispense with real water then look out dry planet.

Other space and time creatures may come to loathe the newly empowered desert flower. With the desert flower in charge, other creatures will become too wet or too hot. Only the desert flower will bloom as the pinnacle of the universe's success.

Bundals wonder what stories the desert flower will someday make up to prove it has created a better universe.

What stories will you make up to convince yourselves of human progress?

It is one thing to fight over made-up stories. Spatial engineering only allows made-up stories.

It is another thing to fight with real stories. Temporal engineering allows for the making of real stories.

Made up stories may help or hurt.

Real stories will help or hurt.

Humans are so rare. You are an extinct species by definition. We have yet to find any other species like you.

Welcome to my recipe for temporal engineering.

Figure 2: Tubal's four dimensional signature in two dimensions.

SHFTSS

SHFTSS is an acronym for

- Seeing
- Hearing
- Feeling
- Tasting
- Smelling
- Sensing

Bundals and humans do all of these, and more.

SHFTSS makes a listener work.

> Readers do not like to work unless they are learning something within just a few seconds.

Whenever you encounter the term, replace it with the phrase, "asynchronous, multi-sensory perception."

> Phenomenologically the perception is isochronous.

> Ontologically the perception is anisochronous.

SHFTSS is used to prevent a neologism such as all-sense.

> Is a SHFTSS a neolexia?

SHFTSS is grok without loss of identity.

The composers of this recipe felt it necessary to introduce multiple forms of sensory perception when eating physics. Humans tend to use vision as their primary form of communication. "Did you see the concert," is a question few humans would find strange. Bundals thought humans listened to concerts.

The Bundals main focus on sensation involves taste. Human babies, yet to be trained by other humans, have a primordial inclination to using taste as well. The mother's womb is experienced by taste. A human babies' first taste of air shocks the baby into crying.

If you believe a baby wants to suck air, just put a mother's nipple close to its mouth. Once the baby learns the difference, will it choose the nipple or the air? Which tastes best?

Philosophy must feel right before we believe it. Philosophy must taste right before we share it.

If a Bundal was forced to lose all of their senses but one, they would choose to retain the sense of taste. This is the primordial sense necessary to navigate the universe.

To use a human term: This will be easily seen when we get into primordial physics.

To use a Bundal term: This will be easily SHFTSS when we get into primordial physics.

Bundals describe reality on the basis of two, temporal dimensions. Humans have historically imagined only one temporal dimension. As a result, Bundals speak of the senses as both input and output. Humans tend to speak of the senses as mostly input.

The human preoccupation with measurement is an artifact of being mono-temporal interpreters of reality. They use coordinate systems in place of bi-temporal knowledge.

Bundals are well aware of their bi-temporal location. They know all senses should function as inputs and outputs.

Talking and singing are Bundal senses.

Humans do not think of talking or singing as a sense. Output, what affects others, is of secondary concern to humans. Humans focus on what affects them.

Some humans even doubt a tree makes a noise as it falls in the forest unless they are there to hear it.

Humans are inconsistent. Deep down they have to know that SHFTSS is as much concerned with output as it is input.

> Bundals are confused as to why humans see smelling primarily as an input. Other earth animals clearly understand it to be an output. Humans should know smell is an output. Both male and female humans put on perfume to attract animals.

SHFTSS reminds humans the universe is more than sight.

> Astronauts, cosmonauts, and taikonauts all claim space has a smell. They claim it smells like burnt gunpowder. This is either a waste of spent rocket fuel or the big bang really burned things up.

SHFTSS serves in this composition as a noun and as a verb. When used as a noun it is acceptable to imagine any one of the human or Bundal senses. When used as a verb, just imagine the word as SHFTSS-ing. An alternate translation might be seeing, hearing, feeling, tasting, smelling, or sensing, without giving priority to any one of the senses.

Physics cannot be relegated to space and time. It cannot survive within mathematics. It is not what we see. It is what we SHFTSS.

Tubal's recipe for physics involves all of the senses. The final dish will enable humans to SHFTSS two temporal dimensions and two spatial dimensions.

Too Much Space

Reserve for Bundal Log Identification

We do not need more space. We need more time.

How much space does it take to hold one dimension of time?

This is a Hermann grid in two dimensions. Examine it closely. Do you not see the dots coming into and out of existence? If your brain sees them, are they unreal?

Maybe the use of more than one sense would help you SHFTSS whether the dots were real or not.

Figure 3: A Hermann grid.

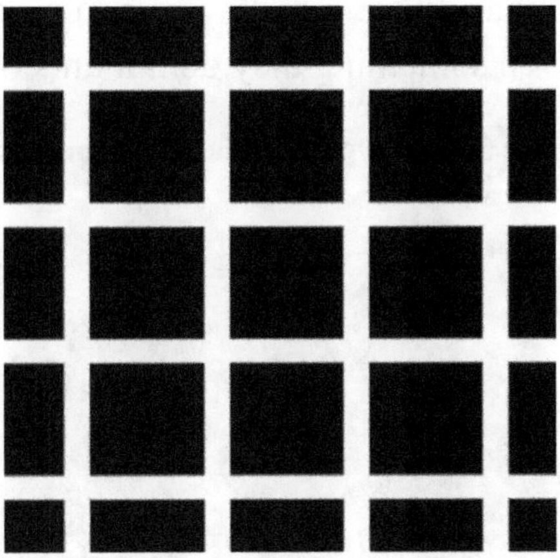

Now let us add Calabi-Yau manifolds to our grid. These are supposed to represent six additional dimensions. This type of model is used in many physics textbooks. Our model now has eight dimensions with the annoying dots popping into and out of existence removed.

Figure 4: A Hermann grid with Calabi-Yau manifolds.

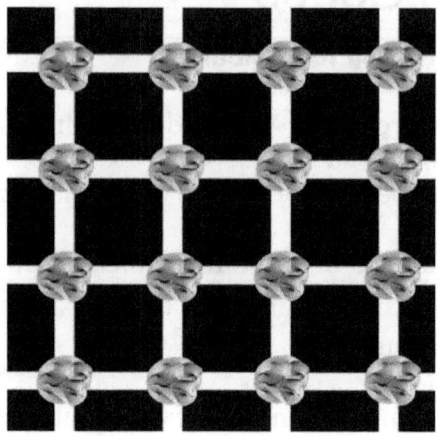

Humans are so funny. Their eye sight fools them yet they insist on seeing more. Human readers will want to see a close-up of a Calabi-Yau manifold. Humans have a feeling of assurance when they can see something. Even something they do not understand.

Figure 5: One type of Calabi-Yau manifold.

Humans cannot see Calabi-Yau manifolds. They see a computer generated graphic of a mathematical equation used to produce a piece of art. Enjoy the art.

This computer generated graphic is best described as a fiat Calabi-Yau manifold. It is a Calabi-Yau manifold because humans say it is a Calabi-Yau manifold.

Humans set up the definitions. Humans built the computers. Humans created the equations. Humans programmed the equations. Humans assigned the colors to different parts of the output. Humans printed the final result.

Humans are now confident they have explained reality with hidden dimensions. Bundals are confused. Science is to be trusted since it is supposedly objective. Religion is rejected since it is composed from fanciful, made-up human stories.

> Humans use fiat money all of the time. The value of fiat money is directly correlated to the strength of the military of the country which backs it.

> The value of the Calabi-Yau is directly correlated to the strength of the funding dollars which back it.

Is a Calabi-Yau manifold art or science? Religion allows a place for both.

We Need More Time
Reserve for Bundal Log Identification

The more dimensions mathematicians add, the less annoyed physicists become. The number of possible universes for humans to explore is simply exponential.

Multiple spatial dimensions above three are certainly easy on a human's eyes. They cannot see them. The greatest progress in science is made when you are observing what you cannot see.

The human race has a history of adding more and more spatial dimensions to reality. How many spatial dimensions are there?

It depends upon the time you ask.

It depends upon what world you live in.

It depends upon your theory.

Whatever happened to time? Why has it always been stuck as a single dimension? Sure, some humans believe we can travel back and forth into time. But it is still just one dimension.

Figure 6: Poor lonely time.

This composition seeks to enfold the temporal orphan within the tapestry of creation. Bundals will give humans a sibling dimension. This composition is a recipe for two temporal dimensions.

Credits

No Reserve for Bundal Log Identification

The Hermann Grid comes to us from http://en.wikipedia.org/wiki/File:HermannGrid.gif on human time July 19, 2012. It was authored by en:User:Famousdog. This is a file from the Wikimedia Commons.

The Calabi-Yau Manifold comes to us from http://en.wikipedia.org/wiki/Image:Calabi-Yau.png on human time July 19, 2012. It was authored by en:User:Lunch. This is a file from the Wikimedia Commons. This file is referenced under the Creative Commons Attribution-Share Alike 2.5 Generic license.

Composer's Notes

Prolegomenon, Prelude, and Aftertaste

Every attempt has been made to compose the information within the Bundal log into a form most humans can understand. The composers have attempted to create a consistent format to aid linearly thinking humans to access the non-linear, bi-temporal thinking of the Bundals.

This is done through the use of formatting and op-narrative.

Formatting

Formatting is designed for twentieth-century electronic communication. As a primitive science, still in its HTML 5.x infancy, the final product will vary depending upon the device used to access the data. This is acceptable since twentieth-first century humans love variety over uniformity.

> Those who value uniformity may read the Bundal Log in Latin or in a book.

The following descriptions are reflective of the Author's desires and may have no correlation to the final product you read.

Major and minor headings have a consistent format throughout this work.

Required information is in left justified format. Please enjoy reading this type of information.

> Indented information is of a different quality.

Indented information may seem comical or cynical.

It usually offers a more complex composition of pertinent issues.

It may seem to stray from the subject without dishonoring the hermeneutic. This is an artifact of Bundal reasoning.

Bundals have multiple perspectives as a result of multi-temporal awareness. Humans may think of it as if the same subject were being treated by a child and an adult at the same time, even though it is the same Bundal. Sometimes it reads like a child. Other times it reads like an adult. The maturity level or quality of insight varies greatly within the same block of material.

The child level offers the greatest insight.

The adult level is often complex.

There is repetition throughout the Bundal log. This is an important aspect of music and thought. Do not deny the necessity for repetition.

The complexities of the adult level will prove frustrating to those who SHFTSS this composition.

SHFTSS is an acronym for the six classical senses of human beings. Humans see, hear, feel, taste, smell, and sense (SHFTSS) their environment.

Inter-species dialog, such as that between Bundals and humans, is grossly misunderstood if one particular sense is deemed dominate over all of the others.

Humans tend to favor sight. Bundals tend to favor taste. You can imagine how incoherent this composition would sound to humans if every time we talked of measuring space we would describe its taste rather than a visual measurement on a ruler.

We now demonstrate an additional formatting level within our composition: The deeper indent. The deeper indent is used to alert the reader to necessary thoughts for further contemplation of issues raised by the composition. For instance …

There is a taste to space. Space can be measured by taste. It may be an acquired taste for humans.

Humans can tell the difference between eating a ten centimeter concrete block as compared to eating a twenty centimeter concrete block. The taste of the last ten centimeters of the twenty centimeter block would be flatter than the taste of the first ten centimeters. A one-half centimeter metallic rod tastes stronger than a one-fourth centimeter metallic rod. The difference in taste is a measurable quantity and quality.

Some humans think that all measuring devices are the same. One ruler is as good as another. Scientifically inclined humans know this is not true.

The length of every ruler depends upon pressure, temperature, comparative velocity, gravitational proximity, the refraction of light and other parameters. No two rulers are exact, even if they contain the same number of atoms.

It may be possible for scientifically inclined humans to SHFTSS the possibility that taste may offer a more exact measurement than sight.

Bundals believe humans have an unrealized proclivity towards taste.

They remind us our entire advancement as a species has depended upon food. We have continuously manipulated space and time to produce good taste. Alchemy developed through smell and taste. Humans claim there is a sweet taste to success. Humans claim there is a bitter taste to defeat. Bundals point out that most human prayers are prayed just before tasting.

Human babies know not to trust the initial illusions of developing sight. They verify reality by putting objects into their mouths.

The maternal relationship has a foundation based on taste. Societies that use bottles have already trained their offspring to love technology.

Bundals believe our one hope as an advanced species is due to the fact we have the culinary sciences. While this may insult the many fans of the hard core sciences, the Bundals' aversion to numbers means they give greater credence to pie than to pi.

Cultures make peace through the sharing of meals. Bundals want you to know the round biscuits we placed on our first deep-space satellites tasted delicious. Humans called them Voyagers and each one contained a disc from earth.

All species know the universal taste for space and time is the delicate combining of diamagnetic and paramagnetic materials. Human chefs did an exceptional job with the only complaint being the disks were salted a little too much with uranium-238.

Composer Joke: Sagan the pioneer should have used an eight-track tape instead of a record. Bundals like spaghetti.

Taste is important, even to human beings.

The composers felt giving priority to any one sensory type does not do justice to other species who have differing forms of sensory apparatus. The best we can do is to use the acronym SHFTSS as a way to accommodate human interpretations of reality. Our composition is then able to take a narrative form somewhat appreciable by human beings. The use of the acronym serves as an irritating reminder that the Bundal log is being redacted for human consumption.

Op-Narrative
Reserve for Bundal Log Identification

Multi-temporal awareness allows Bundals to view reality from the perspective of multiple disciplines, at least as defined by humans. It is as if their senses incorporate elements of history, geography, typology, physics, chemistry, theology, mathematics, music, thermodynamics, social science, psychology, art, and fiction all rolled into the same material. When composing their view of reality into a human information system, it seems as if the final composition is filled with sentences that continually change the subject. This aspect of the composition accurately reflects the fact that Bundals hold differing perspectives at the same multiple times.

The composers have addressed this issue by sorting out the fact that Bundals are aware of at least two, separate temporal dimensions. We therefore will simplify their various perspectives into a sort of human yin and yang. A yin and yang may seem like opposites to humans. Bundals see them as two sides of the same coin; easily discerned through the two, separate temporal dimensions.

We are aware that specialists of particular human disciplines will grow weary with materials from other disciplines. Readers unacquainted with the history of many of the human disciplines will wonder "Why was that material included in this composition?" It was included because Bundals do not have the same categories of information systems as humans. The simplest

30

example is Bundals do not differentiate between science and art. Their understanding of religion is basic to their ontology.

> The final translation of the Bundal log is divided into three separate compositions. The three compositions give priority to post-enlightment sensibilities. It artificially divides the Bundal log into physics, religion, and ethics. This division is our attempt to accommodate human sensibilities as they exist in the twenty-first century. This book represents the first division, that of physics.

> Unfortunately, we would do both species a grave disservice if we tried to completely separate the Bundal log into various human disciplines. The Bundal log will change categories from time to time without warning.

If this composition grows too confusing to the reader, then just skip sections. Bundal thinking is not linear. Humans can get frustrated quickly when they do not see verbal connections logically laid out for them. When in doubt, just skip to the next section.

Human representations of Bundal thought look like op-art and op-narrative combined with dodecaphonic percussion music, perfumed exhaust smells, sensual barbs in the skin, all coupled with sweet-sour tastes. Skipping sections or focusing on just one aspect of this composition will sooner or later return you to a perspective you enjoy.

> Bundals are justly proud of the indented sections, especially when humans have to skip them or do not understand them.

There are numerous other compositional problems briefly touched upon here.

- Humans have two language systems, mathematical and narrative. Bundals have only one language system.

31

- Humans interpret information logically and emotionally. Bundals interpret information holistically.
- Humans describe facts as either true or false. Bundals hold facts in a sort of superposition. They resolve the facts into true, false, or indeterminate.
- The Bundal propensity for holding information in superposition means it looks simultaneously ambiguous and repetitious to human beings.
- Bundals cannot differentiate between fiction and non-fiction. Humans cannot differentiate between fiction and non-fiction.
- Bulleted items are not to be skipped. They are lists, meant to be read.
- Bulleted items do not format well on twentieth-century electronic devices.
- That which is most important is often treated with the least respect. Maybe a book form is better after all.

Humans categorize information. The way Bundals categorize information is described elsewhere.

Human categorize as a way to describe specific parts of the whole universe. Each category circumscribes an academic discipline or human practice. Each category is made up of memories enabled by books, music, art, articles, videos, tribal allegiances, smells, tastes, feelings, rituals, and imagination. When you put the parts of any one category together you do not have a whole. You only have one category. It takes the combination of all of the categories to have a whole.

The problem with human endeavor is that practitioners within each category mistake the parts of their category as representing the whole. They fail to realize the parts of any one category only represent a completed human category, not the whole universe.

Each category is a half-truth needing to be integrated with the whole.

Half-truths are always misleading. Cacophonies hurt the ears while symphonies allow each instrument to be heard.

This makes science seem like a truth onto itself. This makes religion seem like a truth onto itself. Half-truths and categorical truths are indistinguishable.

Compare two separate half-truths and both seem to be correct while disagreeing with each other.

Bundal Joke:

Question: What is the difference between quantum mechanics, relativity, evolution, and thermodynamics?

Answer: Really nothing.

Every human category is needed to be able to properly understand a Bundal's holistic perspective of reality.

Even so-called human insanity offers a nuanced view into parts of reality. All so-called human disabilities can offer perspectives into reality seldom appreciated by other humans.

Bundal information is communicated through more dimensions than humans can recognize. Even on earth, some animals can see, hear, taste, smell, and feel portions of reality outside of the human range of sensitivity. Bundals go beyond known earth animals and plants in being able to sense two temporal dimensions rather than one. Humans know there are two directions of time, backwards and forwards, but seem unable to differentiate the two, actual temporal dimensions.

Humans are becoming aware they are no longer the center of the universe. God is. The human disciplines are far too anthropological. They fail to recognize their limited perspectives. They fail to recognize the limits of human perspectives. Only vain imaginations can enable a sinful species to approach God.

Bundals know that humans will eventually vomit over their limited sensory abilities. Humans will engineer themselves into new organisms capable of expansive extra-terrestrial travel and intensive intra-terrestrial pleasure.

Pornography may become a religion. Hungry humans will super-size their SHFTSS.

Dimensional Language

Composing the Bundal log is complicated by the place of words in communicating Bundal versus human information.

- Bundals communicate primarily through bodily expression, including all of the typical human senses. They fail to be self-conscious of the importance of words.
- Humans communicate primarily through words. They fail to be self-conscious of the importance of bodily expression.
- Humans tend to understand sensory production in terms of input rather than output. Human smell is both an input and an output. Hearing is a sound input. Humans do not normally categorize talking, singing, and making other noises as a sense.
- Humans, recognizing only one temporal dimension, give priority to sensory input. Bundals, recognizing two temporal dimensions, understand sensory qualities as both the inputs and outputs of the living organism. Both are needed to properly understand sense.
- Humans have run up against the difference between sensory input and sensory output as they struggle to understand the measurement problem within Quantum Mechanics. What is objective output verses subjective input?
- Ethics understands that input and output are holistically related. Your pain should not be my pleasure. Darwin unduly limited reality to survival of the fittest as will be SHFTSS-ed in other compositions.

Bundals do not use books. The best we can describe what they do is their use of apparent four-dimensional artifacts capable of communicating sensed portions of reality. It seems holographic from a human perspective as long as holography is understood within a SHFTSS context.

It seems that words will take on an ever decreasing significance for human discourse as we evolve in our use of electro-magnetic communication. Some twentieth-century humans already sense this evolution.

Bundals wonder why humans think words are so important. They ask how we explain that the number of standardized words used in email exchanges declines rather than expands with usage. One would think the lack of bodily expression creates a need for an expanded vocabulary?

Bundals believe humans may become more Bundal-like as our expertise in electronic communication evolves. They even believe we will cease to use complete words allowing only a few letters to substitute for complete bodily expressions. LOL

Bundal language is limited to one noun and two adjectives for each object it describes. Bundal language is limited to one verb and two gerunds for each event it describes. There is never a noun without a verb. There is never a verb without a noun.

It seems as if the human notion of numbers is naturally incorporated into their narrative account of reality. Our sense of numbers seem as a fiction to Bundals. Their numbers can function as adjectives, nouns, gerunds, or verbs depending upon the narrative structure. Ours seem related to equations or narrative nonsense as will be SHFTSS-ed elsewhere.

It must be remembered, throughout this recipe, that we are using that which is least meaningful to Bundals as a way to connect to humans. We are using words.

If only we were sea captains logging about our conquest of new lands. We would at least share a similar world view in describing the new world. Songs would be sung about our exploits. New foods would be introduced to our diets. We would live an exciting new adventure.

We can explain the peculiar use of Bundal nouns and verbs in the following figures.

We will pretend that the visible color spectrum for humans is of interest to the Bundals. (It is not.)

We will temporarily pretend, for the sake of mere example, that black and white form two species of a dimension we will call the gray-dimension.

String theorists would investigate G-strings.

As a substance, so-to-speak, ~~we~~ Bundals would use gray as a noun. As a noun it is modified by the two adjectives, white and black.

We will pretend that graying describes the process of the gray-dimension. As a process ~~we~~ Bundals would use graying as a verb. As a verb it is modified by the two gerunds, whiting and blacking.

Pretend there are sixteen gray colors. As nouns, each color sits on a coarse-grained continuum running from white to black. As verbs, each color sits on a coarse-grained continuum running from whiting to blacking.

Figure 7: Identifying colors with course granularity.

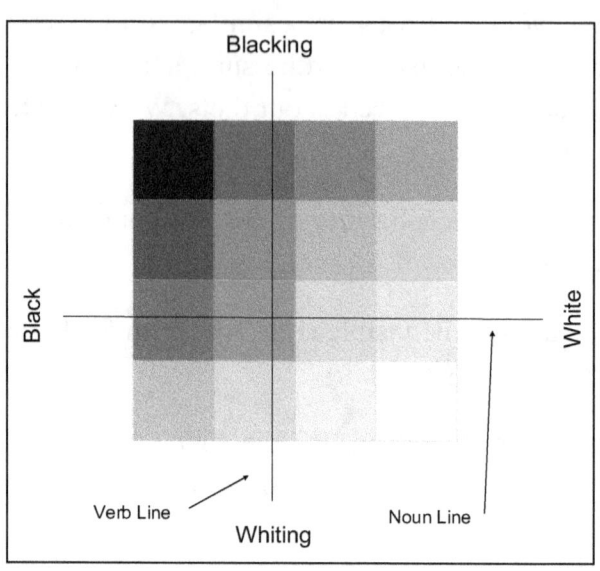

Think of this figure as a Bundal noun line intersecting a Bundal verb line to describe a unique gray. The adjectives used for the noun will shift the verb line right or left. The gerunds used for the verb will shift the noun line up or down. Each Bundal sentence identifies a unique gray.

The same Bundal noun and verb can describe all sixteen shades of gray.

Humans try to name each gray with a unique word.

Which one is charcoal gray? Which one is gainsboro?

How many names for gray can you find on the planet earth? How many colors of gray are there on the planet earth?

The differences between Bundals and humans are stark. Humans tend to quantify the amount of white and black in each gray noun. Humans say the gray has a certain percentage of white and a certain percentage of black. Humans see the white and black adjectives as intrinsic to the noun.

Bundals understand that the noun is defined by the integrative process or verb. What the noun is doing defines its level of gray. The verb line, defining the blackness and whiteness of the noun, moves right or left. The whiteness and blackness of the gray are not intrinsic to the noun. The whiteness and blackness take on meaning only when there is a process involved (as described by the verb) to produce the final white-black gray.

Humans tend to quantify the amount of energy or input and output over time in regards to measuring the amount of whiting and blacking occurring. Again, they are misunderstanding that blacking and whiting, as gerunds, are not intrinsic to the process of graying. The process involves a substance. There has to be a noun involved in the process for the process to have meaning.

There is a greater *faux paux* as well in that humans measure time in spatial terms rather than temporal terms. This *faux paux* is described elsewhere.

Bundals understand that whiting and blacking require the noun in order to have definition. The noun line shifts up or down based upon what the noun is doing. This noun-shift defines the amount of whiting and blacking occurring within the process.

Bundals would describe the unique place in the gray-dimension as having one color with two adjectives, black and white. What we see as a gradient of spatial colors appear as a temporal phenomenon to Bundals. The Bundals have one verb for this temporal change of colors. There is no equivalent human verb. The closest translation we can come to is that of frequency modulation, although this concept is far too spatial to satisfy a Bundal. The adjectives now become gerunds in Bundal parlance. The gerunds describe the temporal event as blacking or whiting. This ambiguous information seems

39

non-descriptive to us. But for Bundals, the information contained in the contrasting adjectives and noun intersects with the information given in the contrasting gerunds and verb so as to produce an exact location or description of the color. It is as if a noun and a verb come together to create a unique reality or story.

> Whitehead's contrast is Heidegger's unveiling. Plato's shadows create light. The Bundal place slices through human aesthetics and human thought to become Kant's categorical imperative. Kant would disapprove. If Kant knew the Bundal's God, he would approve.

On the other hand we humans come up with names for each of the colors. Each name we use falsely describes an intrinsic quality within reality. What is the difference between charcoal gray and gainsboro?

> The English idiom should probably be restated as, "on the other eye."

The more colors we use in our discourse, the more names we create.

> Bundals believe that human beings should admire the Incas and Chinese among all cultures. The concept of one pictorial symbol for every idea clearly expresses the human experience in using language.

This creates a naming problem for humans. The finer the grading of reality, the more difficulty humans have to describe unique, supposedly intrinsic, colors. Bundals would use the same language and adequately describe the unique color place regardless of fine versus coarse granularity.

Figure 8: Identifying colors with fine granularity.

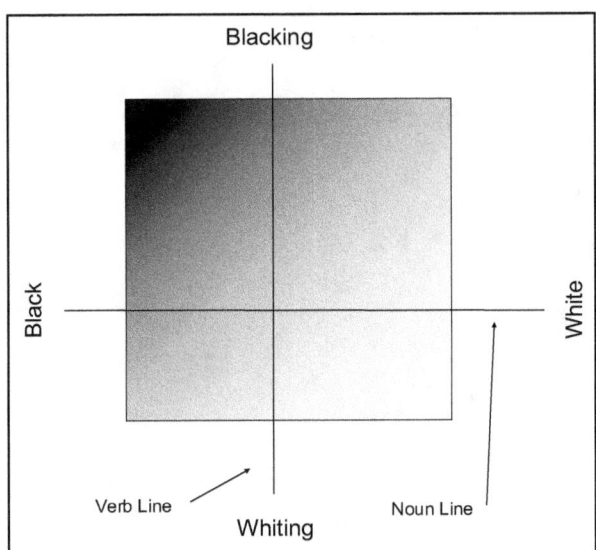

Now how many colors of gray would a human need to identify unique places in the gray-dimension? (Or is it grey?) The Bundals have no difficulty in using complete sentences to describe unique places in their stories.

We have been speaking of gray but Bundals know complementary colors, so pleasing to human eyes, combine to form human grays. Even humans know white light can be separated into the colors of the rainbow (and beyond). It turns out the gray-dimension even contains colors, at least by human standards.

Bundals are not confused but humans could be.

Figure 9: The gray-dimension in color.

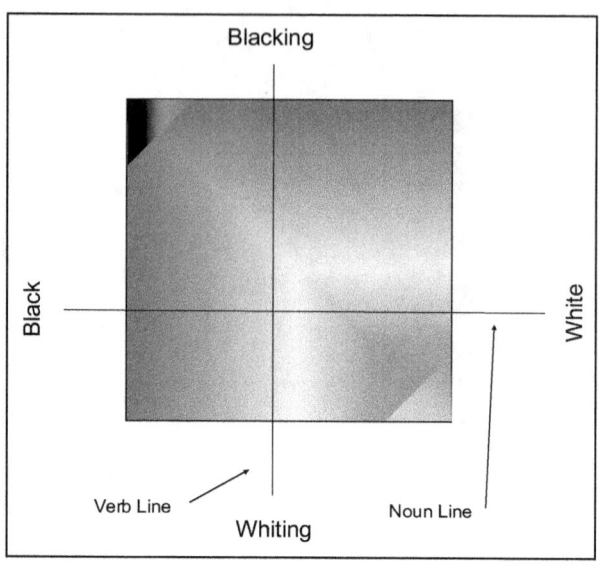

(Figure 9 is in color in the original and electronic editions.
It prints as gray in book form.)

The foregoing section should be read with Anne Murray's rendition of *I Can See Clearly Now* streaming in the background.

You should be quite convinced that humans will never completely understand Bundals. Nor will Bundals ever completely understand humans.

Bundals will learn from our blindness.

We will learn from their sour taste.

Other issues of translation that will come up as you read this composition. The composers remain committed to the task of human communication. An esoteric work helps no one.

Let the music begin, even if parts taste sour.

Hungry?

When as a reader will you get to the central idea of imagining two temporal dimensions?

Paradigm shifts require preparation. You do not jump from a red world to a blue world and accurately describe everything from a red perspective.

Changes of perspective have occurred throughout human history.

- 1800's- Solid is solid.
- 1900's- Solid is more empty than full.
- 2000's- Solid is mere light.

Tubal's Recipe guarantees you will be able to imagine a second, temporal dimension. Great hunger is needed as this type of imagination requires a wholesale paradigm shift in how humans describe reality. Historically we have seen ourselves living in a three dimensional world traveling through time. Tubal sees it differently.

Tubal starts us on a journey in space and time, but declares space and time are insufficient to describe reality. Rather than pit religion against science Tubal asks what if space and time were not the fundamental dimensions to describe reality. What if gravity and light were the two fundamental dimensions? How would this look? Or feel? Or smell? Or taste? Or sound?

This requires space and time to be derived from gravity and light. This insight contextualizes Einstein's Relativity in terms of gravity molding space and time and explains the apparent constancy of the speed of light from within the space and time framework.

The power of gravity and light would also explain the type of energy necessary to continuously create space and time. This would contextualize Dynamics within a theory of gravity and explain why there are apparent zones of high entropy and low entropy within space and time.

What kind of space and time do gravity and light form? Obviously it will depend upon the characteristics of gravity and light. Tubal describes gravity and light as having two species. Space and time have two species as well. All four species are described in Tubal's recipe.

This means we live in a world with two temporal dimensions and two spatial dimensions. Tubal reinterprets common human experience to explain how we live in such a world. Quantum Mechanics is naturally explained on the basis of our bi-spatial, bi-temporal world. Multiple types of symmetry are derived from this explanation. Tubal's paradigm runs contra against the current need to multiply the number of spatial dimension mathematically necessary to describe reality.

Tubal sees the dimension of light as an information system utilizing gravity to form space and time? Holistic information, at least from a human perspective, explains the need for substance ontology and event ontology as both necessary to describe space and time. Particle-wave duality is derived from this explanation.

Reading this book requires a hungry imagination. It also does not hurt to have a background in Relativity, Quantum Mechanics, Dynamics, German Enlightment Philosophy, and Mathematics. Failing this, you will not get the humor. Having a modicum of background in any of these disciplines will enable you to imagine two temporal dimensions. (Guaranteed)

Is This Really a Recipe?

Answering old questions with alien ideas.

How do you explain an entire composition in one sentence?

Pick one.

- This composition translates an alien log that reappraises three-thousand years of human understanding.
- This composition is a metaphorical introduction to an alien physics.
- This composition is an attempt to enable human beings to imagine two, separate, temporal dimensions.
- This recipe mixes the ingredients of gravity and light to form space and time.

This composition is no more fantastic than what medieval sea captains posted for their king's pleasurable reading. They told stories of sea monsters, kept dairies of raw human emotions and deprivations, wrote details of exciting new discoveries, and described with varying veracity the clear interventions of God on their voyages.

European populations heard stories from the writings of these new world conquerors. Artists drew pictures of sea monsters. Suffering ship crews were made heroes, especially those who did not mutiny. God was praised for all the miracles and the miracles were recounted in numerous barroom conversations.

About the only thing questioned in these reports concerned the new discoveries. Were there really new lands across the ocean? They were probably old lands, rediscovered.

The western hemispherical people who were conquered did not see themselves as living on new land.

History claims they progressed under the conquerors' tutelage which is another way of saying they suffered terribly.

Dickens' *Tale of Two Cities* could just as easily have been the tale of two hemispheres.

Who, what, when, where, and why are all matters of perspective. It is a fact that just because ship-crews saw new lands, ~~that~~ their seeing <u>did not</u> convinced anyone that new lands actually existed.

Fortunately vain imagination won out. People wanted there to be new lands. Only new lands could offer …

- … gold to fill their coffers
- … spices to fill their bellies
- … converts to fill their heavens
- … land to expand their kingdoms
- … adventures to fill their story books
- … a place to escape from what was already known.

And so the new lands came to be in spite of skepticism over reports of what lay over the ocean's edge.

Medieval sea captains sailed in uncharted territory but they at least sailed in our world. They shared a common world view with their fellow citizens. They were not metaphorically hindered from describing their new discoveries. Their stories could directly excite the European imagination.

The common European world view did not protect them against flawed analogies and comical interpretations of reality.

Our favorite sea captain described how thousands of natives attacked his conquistadors. The natives dared to pit spears and arrows against Spanish guns. His log states God intervened. The Spaniards triumphed because so many of the natives were killed by arrows from heaven hitting into the backs of natives who were inching ever closer to the Spaniards.

The arrows from heaven are scientifically verifiable as the story is observable across history in different formats and cultures. The arrows and dead bodies are quite real.

The source of the arrows as from heaven is deduced from logic. Few would believe the Indians would deliberately shoot their own comrades in the back.

We are not so fortunate when we seek to describe an alien physics. The Bundal log presupposes a form of life so alien to us that the notion of mutual understanding is equivalent to a series of redundant paradoxes. Humans have two languages, narrative and mathematical. Bundals have only one language. Humans use words and do not seem to notice how much is communicated through bodily expression. Bundals use bodily expressions and do not seem to realize how much is communicated through words. Fortunately, paradoxes are what make human physics real. So we do have a starting place for our first conversations.

The composers initially grouped Bundal information into subject categories familiar to humans. Unfortunately, linear thinking does not accurately capture bi-temporal experience. This results in a warning for the human reader.

Each of our physical disciplines represents a gross misrepresentation of reality as evaluated by Bundals.

- Quantum mechanics presupposes probability.
- Relativity presupposes determinism.
- Both allow time to flow forwards and backwards.

- Dynamics has a single arrow of time always trending towards higher entropy.
- Evolution trends towards lower entropy.
- All the above theories work quite well within their respective categories.
- Try to combine the above categories and the math goes to pot. And it is not an edible pot.

This does not mean we will jettison human physical insights. It does mean we will need to appropriate them with new understanding. Sea monsters have nothing on Bundals. A new land exists. Few will believe in a new physics. They will believe in everything else along the way. The Bundal logs are worth our SHFTSS.

The medieval sea captains still conquered in spite of denigrating multiple cultures, and violating their own best ethics. Twentieth-century science has also conquered the earth. The major difference between carbon dioxide and a nuclear bomb is the time scale.

Too Many Ingredients Spoil the Recipe
Reserve for Bundal Log Identification

This composition is entitled, *Tubal's Recipe for Physics*. Its claim to fame is its ability to enable humans to imagine a second temporal dimension. There is a converse implication, a temporal yin to the spatial yang. Too many spatial dimension spoil the stew within which we SHFTSS.

The composers recognize most readers will discount the discovery of a new physics. But many will believe in aliens.

This book attempts to do for time what Charles Hinton did for space in his musings on four spatial dimensions. His work was published shortly before the first airplane rides and successfully demonstrated how one might imagine living in four, spatial dimension.

It seems propitious for us to learn a little more about time before we set off on our first 4.3 billion light-year journey to Alpha Centauri.

Space will be aggressively explored when it becomes more pleasurable to do so. After all, the first airplane ride was to a beach.

Twentieth-century arguments about living in orbit versus living on the moon versus living on an asteroid versus living on Mars versus living on Europa will be solved by answering the question: Which place will have the most pleasure?

Bundals tell us that orgasms are possible for humans in a weightless environment. Humans tend to puke during weightlessness. We wonder how this will work out for humans.

What is so important about time?

Human minds and human computers have done a credible job of imagining and modeling universes with increasing numbers of spatial dimensions.

- We can model a three dimensional universe.
- We can model a four dimensional universe.
- We can model a nine, ten, eleven, or sixteen dimensional universes.
- We can model a twenty-six dimensional universe.
- We can model a 10^{126} dimensional universe.

Bundals would disagree with our use of the word credible.

Does the word infinite mean anything to you? Apparently humans can describe infinite multi-verses by modeling no more than 10^{500} numbers of spatial dimensions. This is the supposed number of potential models needing to be examined by string theorists. What is actually real wanders in disguise among one or more of these models.

In a wonderful way human physicists know our universe may be bounded but infinite. Or it may be unbounded yet finite. Other possibilities exist as well.

> This may be a warning to not so easily dismiss the extra-ordinary insights from our Bundal friends.

If all of this is confusing to you, then welcome to human physics.

> Bundals shudder to think what human science would be like if it ever went beyond mere observations.

Bundals wonder why humans have never challenged the need for increasing the number of spatial dimensions. Why not increase the number of temporal dimensions?

> It is at this point Bundals question the invention of human mathematics. If mathematics is universal, if mathematics is a divine language, if math is even more real that the space and time it creates, then why not add increasing numbers of temporal dimensions to our equations?

> A divine language ought to be able to describe 10^{500} different temporal dimensions with as much ease as 10^{500} spatial dimensions.

Transitioning our physics to two temporal dimensions could greatly simplify our understanding of the universe. If we could successfully imagine a second, temporal dimension, then maybe we would not need so many spatial dimensions.

> Armageddon approaches.

This composition sounds Bundal notes for modeling time and space. Their music carries us to an imaginary place where:

- We live within both sides of a two-dimensional space with both sides being temporally separated.
- We live within both time zones of a two-dimensional time with both zones spatially separated.
- The two dimensions, space and time, have two species each. Using human numbers this allows four possibilities for SHFTSS-ing. This simplifies how time and space inter-relate and gives rise to understanding the Einstein-god.
- The two dimensions, space and time, are described as quantum mechanical phenomenon. This aspect of our composition lies so close to us we might as well be examining superposition.
- Examining the Einstein-god will demonstrate the human limitations for SHFTSS-ing reality.
- The quantum mechanical look is actually more real for us humans than three-thousand years of human interpretation would lead us to believe. This composition will parse our history.
- The Bundal model presents integrative answers for the concerns of the Einstein-god who lives within a quantum mechanical universe. Like all models, it is imaginary before it is potential or actual.
- We live within a bi-temporal, bi-spatial universe.

Our recipe demands only the essential ingredients are needed for the best tasting universe. We will now tune our instruments to hear do-decaphonic percussion.

Food Poisoning
Reserve for Bundal Log Identification

Ethics require we start with noting the difference between spatial engineering and temporal engineering. Ethics demand we warn human beings about the social implications from being able to imagine a second, temporal dimension. Digesting such an understanding could release virulent diseases that will wipe out wholesale numbers of cultures.

Ancient sea captains and their syphilitic crews have nothing on those who can imagine a second temporal dimension.

Those who are immoral or altruistic may skip to the sections that deal directly with imagining two, temporal dimensions. All others should SHFTSS the following warning before seeking to understand an alien physics.

These warnings include the Bundal's observations concerning the differences between spatial engineering and temporal engineering. These warnings are sounded between the first and last movement of our composition.

If only Einstein had a warning before releasing his insight on the relationship between energy and mass. $E=mc^2$ SHFTSS like Nagasaki and Hiroshima.

Warning: Like Gods to God-like

This chapter requires audio streaming of dodecaphonic music as it is being read.

Spatial Engineering

Bundals see humans as spatial engineers. This form of engineering Bundals use to create areas of high entropy and is considered of limited value. Bundals prefer temporal engineering. Temporal engineering creates new realities with existences of their own.

Human engineering is constitutionally designed to create objects that decay. Humans claim not to want to do this. Yet they insist on designing objects that decay.

> Human engineers would claim they have insufficient knowledge to create objects "that will last forever."

So much for being like God, in spite of what technology promises.

The Bundal's evaluation of human motives convinces them that human engineers actually prize high entropy creation. The Bundal log contains these two ingredients that do not taste good to Bundals.

- Humans claim atomic weaponry as their highest achievement for peace.
- Human engineers design products that wear out frequently so as to force other humans to buy the product again. This is particularly puzzling to the Bundals. They SHFTSS humans as creating items of real value, then limiting the product's

usefulness so as to secure fiat money. Bundals know the real value of fiat money is as energy insulation or to start camp fires. Fiat money is laughable as far as Bundals are concerned. Birds use it to build nests. Humans who value degeneration love fiat money most of all.

Bundals feel sorry for humans at many levels.

The words played above sound like peace played on an instrument of war. The piece is further complicated when the imagined value of fiat money goes vibrato. The sounds of earth, food, and water are degraded as valueless instruments of brawn rather than creations of superior intellect. The tones of poverty follow as a tremolo to the vibrato of human-made money. These two movements and their instruments form a complementary refrain best interpreted as high entropy. The final result sounds much like electromagnetic waves at 2.72 degrees Kelvin.

Twentieth-century economists focused on debates within their particular discipline. They failed to discern the macro lesson of a flat world. Forget stock markets and GDP. The value of fiat money is directly correlated to perceived military strength. Genghis Khan still lives.

The human disposition towards high entropy is nowhere more evident than their desire to live in outer space. Where do they think they live anyway? Humans inhabit the best space ship ever created for liquid life. What is so appealing about shooting little parts off of their space ship to someday live in high entropy zones?

Spatial engineering irretrievably leads to high entropy. Why the latest and the greatest are seen as progress, rather than as new toys for the trash-heap, remains a mystery to Bundals.

Spatial engineering is enabled through the human ability to calculate effects through more than one spatial dimension of reality. One dimensional engineers do not engineer one dimensional space. You need at least two spaces, one to work on and one to work from. The foundational axiom for engineering is the requirement for more than one dimension whether it is spatial or temporal.

This axiom is never allowed by Euclid with his non-dimensional points and therefore has yet to be discovered by humans.

Temporal engineering, never tasted by humans, requires a minimum of two, temporal dimensions to form a space that produces regions of low entropy. Temporal engineers create.

Spatial engineers aim towards 2.72 degrees Kelvin and lower.

Temporal engineering requires the ability to imagine more than one temporal dimension. Once time is SHFTSS as bi-dimensional, the manipulation of time is accessed through gravity and light.

Tubal has suggested to the composers that we should interject: A better way of saying this is to say that the manipulation of time will be accessed through its gravitational shape. Light, both white and black, is the prime tool for manipulating gravity. This runs contra to current human interpretation.

Light is the constant of the universe. Why look for ontological foundations elsewhere?

The composers fear their current insights depend upon twentieth-century language and may not be applicable in later centuries as the classical views of dimensionality are improved. They therefore stay with the Bundal assertion of white and black light rather than luminous matter and energy and dark matter and dark energy.

The dodecaphonic music continues to play.

Spatial Gods
Reserve for Bundal Log Identification

Twentieth-century jungle movies showed the white man flicking a flame into existence with the use of a cigarette lighter. Native jungle dwellers were awed. The white people were like gods. They could create fire with their magic.

> The fire was created through the talents of spatial engineering. A wick, flint, and flammable liquid came together and were ignited by the input of thumb energy.

Especially poignant for the jungle people was the camera. This device allowed the white people to capture their souls. White people were like gods.

> The pictures were created through the talents of spatial engineering. No one wanted to drop their camera out of the canoe into the river water. So maybe the gods did not really make it.

Once humans learn how to do temporal engineering, they will be God-like, not god wanna-be's.

Bundals understand there will be three epochs of the human race.

- Before the time of spatial engineering, humans prayed to their gods.
- During the time of spatial engineering, humans did not believe in God.
- After the time of temporal engineering, humans will pray for there to be a God.

The real God will seem like a saint once human temporal engineers take over.

Temporal Engineering
Reserve for Bundal Log Identification

Spatially focused minds can write stories filled with temporal para-doxes.

Temporally focused minds can create real stories in space.

Temporal engineering allows the classical measures of story-telling to be abrogated. Story lines, character development, and plots have multiple juxtapositions whose relationships are no longer necessarily continuous. Our era is ripe for discontinuous story telling. Change our time frame and we change. Imagine what will happen when we change two time frames?

No one who studies history doubts that the best of human history is discontinuous. Even the longest stories are discontinuous. In one era, tribal loyalties drive history. In another era, the lack of resources drives history. Yet another era looks to finance as the root cause of the emerging story. Then there are eras motivated by jealousy, intrigue, power, and fame. Other eras are formed by the self-gratification of a few leaders. Religion and idealism drive other eras. It seems as if every part of history has its own story. And we seem quite unable to connect them in a consistent or orderly way.

Discontinuity of history is an artifact of the notion that only the survivors live to tell the tales.

Discontinuity of history is an artifact of linear thinking. Reality must be bi-temporal or it is only a fiction.

War and Sex in the Bi-Temporal City

War in the Temporal City

If you think human warfare has been fierce when people have to die, just wait until we can wage war with no casualties.

The pagan gods were always fighting. It is so difficult to kill a god. Without death, war can last a very long time.

The world needs oil. Countries will fight over oil.

This will be our future.

The temporal engineers knew they could have created cesspools of oil within the bounds of their own country. But their prime directive was to leave things as close to the way it was before, whatever that meant. Legal interpretations, coupled with self-need, allowed the prime directive to be interpreted in many ways.

The Prime Directive of Star Trek fame was predicated upon the belief advanced species could not really help less-advanced species. Of-course, it was acceptable for humans to advance themselves by stealing technology from more advanced civilizations. Twentieth-century law makers certainly thought it beneath their dignity to help lesser species survive.

The temporal engineers went to work. Construction was made to form an alternate reality over the recalcitrant country that would not share their oil. Oil hungry armies were trained.

The temporal engineers ran through their rehearsals. An alternative spatial reality was formed over the training grounds. Every animal organism, within the alternative reality, became like big gooey balls of rolling flesh. There were no bones.

One would have thought this would have smothered all animal life. No. This was an alternative reality. It was as real as if they had taken one hundred and fifty billion years of evolution to form. The animals were alive. They could make sounds. They could roll. They could not chew food. They could not get water. They could not talk. They could not easily lift objects with their boneless hands and feet. They could live within their alternative environment.

The soldiers in training were terrified within their newly created training camp. Leaders, who had fought within other alternate realities previously, rolled over to calm nervous newbies. Many times their rolling had an opposite effect.

It would be hours before the new army realized it was not going to die.

Tools for boneless people had been created and left in the area of the alternate reality. Transportation vehicles had ramps which bulbous globs of flesh could roll up. Weapons required sloshing in order to aim and fire. All of these had been made available before the alternative reality was created.

No one could use such tools in our world. They were perfect for the new world.

After training, all of the boneless soldiers could use them.

Communication was enabled by artificial devices being inserted in what had been the mouths of the boneless people. These

devices resembled prosthetic jaws but were much more complicated.

The field training event went well. Experience grew as to how
to use the weapons. Confidence in life was regained. They even
had ways to feed and drink while in bulbous form. A sense of
superiority over organisms with bones began to creep in.

The alternative reality was reversed. Animals took on the form
of a relieved group. Members of the newly-trained army started
to celebrate.

Many more days of preparation awaited them. They would enter
the new reality then reverse. It only depended upon the temporal
generals as to when the alternate reality was turned on or off.

They practiced and mastered the environment of their new reality.

The oil-drenched country had no idea what hit them. They had been
given an ultimatum to sell their oil to the rest of the world as a gen-
erous price. Failing that, they would be invaded and their assets
seized for the good of the world.

When the alternate reality was engaged over their country, screams
should have been heard. In their stead was a constant gurgle, that
changed in intensity and pitch as the number of terrified choir members joined in or left the choir at varying rates.

The entire country was filled with giant, red, swollen eyes.

Their spatial engineering was not designed for this temporal interface.

Machines with ramps rolled through the country side. Weird looking weapons and tools were seen. Soon there was a foreign soldier
on every corner. Military bases, seats of government, utilities, major

manufacturing plants, food and water resources, and countless other parts of the new reality were now under the control of a freakish, foreign army.

Boneless babies were born during the night and did quite well. No one knew how to feed them. Mothers even questioned whether they were babies.

Mechanical equipment, run by foreign army regulars, picked up existing arms and munitions from their storage lockers. Planes, missiles, ships, tanks, and other assorted military vehicles were destroyed.

The different reality was even run on a different time.

It was all over within an hour, depending upon your frame of reference.

The alternative reality was turned off. Bones came back into place. Animals and humans rejoiced. The oil spigots were set loose for the world to drink. No one was hurt. No one was killed.

At least no one was supposed to be hurt or killed. The leaders of the country were never found but there was a series of strange fluid slicks within each of their homes.

No one in that country would ever want to go to war again.

A war without casualties is hell.

Sex in the Temporal City
Reserved for Bundal Log Identification

They needed to settle an outpost near the new earth. Too many of the crew had died before arriving. Too few were left to secure the possibilities of starting a successful colony. No one was going to have a child only to watch it die for lack of resources. Humans were challenging the very limits of habitat insisting they could live anywhere through temporal engineering.

> The remaining leaders met to formulate a strategy. How could they encourage the few remaining colonists to procreate without forcing them to have sex? The history of the human race was replete with varying taboos, morals, and legal strictures concerning the control of sex. Humans could not be forced against their will to have sex, especially meaningful sex. This was almost a religion rather than a law.

> Two strategists resolved the problem before one night had passed on the sparse outpost. Colonists heard the sound of a retreating transportation ship. Left behind was a new reality.

> They were turned inside out.

This new state of affairs did not kill them. In fact it heightened their sensitivity to their environment in all sorts of ways. It was as if they had been created to live like this for all of their lives.

The pain was terrible. Throwing acid on your private parts and setting your faces on fire would be no more painful than the agonies that rocked each of their nervous systems day and night. There was a form of rest available to them, but the nightmares did not permit them a long sleep.

Soon they discovered the answer to their pain. Sex would neutralize their pain. Sex was not pleasurable; it only killed the agony and allowed them to live without crying or screaming. As long as they were having sex, they lived without pain.

The absence of pain was superior to the addition of pleasure they had remembered about sex within their prior reality. They did not even need to care for the other person. Their only focus was to relieve their own pain.

There was another way to avoid the pain as well. Some sort of spatial relationship was established through the sex act. As long as the females were pregnant, both father and mother lived pain free. This nirvana would continue for about six months after the birth of the inside-out children.

Then the process of pain would begin anew for the adults.

Children stayed pain-free until puberty, which occurred relatively early in their lives. Adolescents were capable of vigorous and frequent pain relief.

Sex was nothing compared to procreation. Sex was only a temporary relief. Procreation was equated with salvation.

The inside-out people thought about killing themselves but their inside-out bodies proved quite resilient to damage. Their new reality seemed responsive to rapid healing since the inside was on the outside.

Someday, other humans would come back. They would restore the outpost's reality to something less painful. No human would leave their brothers and sisters to suffer for an eternity.

Unfortunately, pain and procreation worked together to create an insanely normal people zealous for others to experience the glories of agony. The masochistic colony grew in numbers and power. They would build ships that would soon take their superior form of reality to other parts of the universe.

Pain would win.

A new religion would be born based on the right of all humans to relish pain. This new religion saw Jesus as the first-fruit, the elder brother, and the divine example of those who would not only drink his cup; they would create it.

Oh for the days of cigarette lighters and nuclear bombs.

What Will You Do?
Reserve for Bundal Log Identification

When human beings learn how to manipulate reality through temporal engineering, the real God is going to look like a saint.

This composition sings a constant refrain. Human experiments in temporal engineering will have unintended consequences since humans cannot adequately SHFTSS.

The desert flower will feel like a removable cactus thorn in your stomach, eternally present. Eternally present means you are unable to spatially move it.

If you believe using carbon resources is problematic for twentieth-century energy needs, just wait until you start using up temporal resources.

You miss-SHFTSS time as an unlimited energy resource. It is not. It is gravitationally limited.

Humans may take comfort. There is a universe Bundals fear. It is a universe filled with Tabasco Sauce.

You have been warned. If you read further and finally understand how to imagine a second temporal dimension, you will join a cadre of people leading others to god-like powers.

What do your ethics tell you to do? Will you read on? Or will you be satisfied with one dimension of time?

Human History of Spatial Engineering

Humans have been digging into spatial engineering since the discovery of opposable thumbs and large cerebellums.

Dolphins may challenge this.

The human concept of the universe has historically been spatial with the one space evolving through time. A place is a space with a time signature. Space predominates, at least in the classic view.

Some speculated space repeated itself. It would live and die and live again. This space was controlled by time which was seen as cyclical.

Others believed time to be linear and non-repeating. In this case space would change into new forms and never really repeat the old forms.

Can you really step into the same river twice? The river is the same but the water is different. It is always different water molecules flowing past you. As far as humans are concerned, the river-space looks and feels the same, especially on a hot day.

Even Plato's shadows were spatial. They required a cave and a wall. No one asked about the light.

Spatial engineering has an eastern and western history. The composers choose a western motif to describe spatial engineering. The engineering accomplishments of the human race go far beyond mere western motifs. The western motif does provide a semi-linear frame for the op-narrative picture. As such it represents a further condescension of the Bundals in favor of effective human communication.

The semi-linear western motif starts with Euclid's triangle, moves through Hyperspace, the Big Bang, Zero Omega, and concludes with space and time. Even if we had used eastern motifs, readers need to remind themselves all human history is provincial.

Euclid's Triangle
Reserved for Log Identification

Why do humans have such a pre-occupation with space?

In part, it is because we can manipulate space. It is far more difficult to manipulate time.

Human spatial studies started by measuring triangles on flat planes such as the earth. Bundals find this confusing.

Sometimes songs have to be sung twice or we do not hear them.

Human spatial studies started by measuring triangles on flat planes such as the earth. Bundals find this confusing.

> We are reminded not to laugh at the logs of medieval sea captains.

> Thank Allah for seven hundred years of Muslim influence in Spain. Catholic Europeans knew the earth was flat. Muslim scholars knew the earth was round. When King Ferdinand and Queen Isabella dispatched the last enclave of Muslim leadership from Spain in 1492, their culture had known for seven hundred years the earth was round. How ironic that the most Catholic of European leaders were quite assured their financial investment would not sail off the edge of the earth.

> Christopher Columbus was relieved to find a King and a Queen who knew that flat triangles proved a rounded earth.

The composers can understand the Bundals confusion with the human species.

As long as ancient measuring devices were as flat at the earth, the interior angles added up to 180 degrees.

It is always a shock for scientific people to discover the basis of their geometry was founded upon a fiction. There is no such thing as a flat anything anywhere, thanks to gravity.

We then progressed to studying triangles on spheres and hyperbolic saddles. The interior angles of these triangles summed to greater than one-hundred and eighty degrees or less than one-hundred and eighty degrees. It depended on whether you were riding on a ball or on a horse.

Figure 10: Changing interior angles of triangles.

Triangle on Spheres, Flat Earths, and Saddles

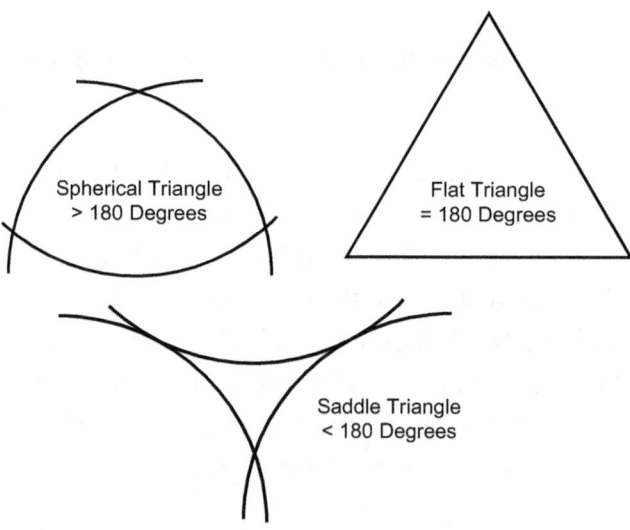

Hyperspace
Reserve for Bundal Log Identification

The Einstein-god changed our conception of space and time creating hyper-space, where space and time are interchangeable. At this point time came to predominate in our understanding of reality. This is especially ironic since classical relativity has only one temporal dimension interchangeable with three spatial dimensions. Einstein's hyper-space was subservient to time even though time was outnumbered three to one. God always sets things up so the mighty fall.

The pictures of giant gravity wells bending the fabric of space and time are illustrated in many science textbooks by giant balls resting on the surfaces of semi-taunt sheets. Each sheet is indented, or curved in the spot on which the ball rests. Some illustrators caution the viewer that it is both space and time that are bent to create the bulge. Unfortunately, this picture gives us a spatial view rather than a temporal view. Gravity wells slow down time. This slowing of time shrinks the apparent size of space. This is opposite the picture of the bulging sheet. The bulge in the sheet makes it look like space is stretched and expanded by the weight of the ball. Actually space is shrinking with time being the predominate factor for the shrinkage. Gravity is best illustrated by shrinking sheets, not balls rolling down into space-time holes.

Figure 11: "Weight" in middle pushing down on orthogonal, two-dimensional fabric. Note the decrease is spatial size.

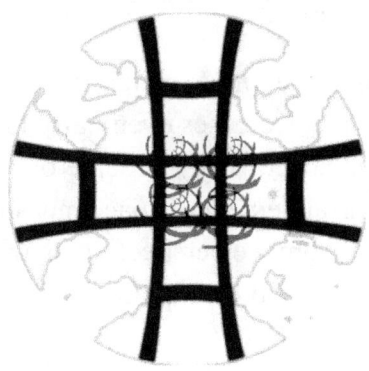

Bundals like to think of shrinking space as pulling other objects towards the place sucking in the space. It would be like riding on the surface of a deflating balloon. Gravity is the feel of surfing on a gravitational wave. This wave is felt by humans as the temporal differentiation between the spatial dimensions.

Temporal differentiation presupposes at least two temporal dimensions

Let us imagine two objects, gravity wells if you please, pulling on each other. This pull between two objects, due to the shrinking of space between them, is offset by an orthogonal, temporal differentiation. The fact that time is slower in the large gravity well, enables the object in the faster time zone to counteract the shrinkage of space. This is characterized in classical physics as centrifugal force. Even a larger object will attempt to temporally fly off from the shrinking effects of space introduced by the smaller object. When temporal differentiation is stable or sufficient, it offsets the effects of shrinking space and the objects stay consistently apart.

One may also conceive two objects having internal metronomes marking their pace of time. The Einstein-god can hear both metronomes. If the difference in rate, between the two metronomes grows the objects will surf farther and farther apart regardless of how much space shrinks.

If the metronomes begin to match each other, the temporal, counter-acting force decreases eventually failing to differentiate the two objects in space. They become one. This is known as a collision, a new arrangement of space measured as a past calamity in time zone B from the perspective of a future reconstruction in time zone A. This will be explained later.

It is ironic that colliding chaos is formed in close to synchronicity temporal conditions. Spontaneous organization of space is formed in far from equilibrium conditions.

Black holes operate on the same principles. Here inside-space is united as future-time contracts. The formally separated objects become temporally undifferentiated. Since inside-space is two-dimensional, as sung elsewhere, the final effect is surface SHFTSS versus volume SHFTSS since volume is only produced by temporal differentiation. (Explained later.) Also note that wet gravitational fluid evaporates, releasing cold.

When two dimensions of time are introduced a ready explanation is made for the fact that there are two foci rather than one to measure the surface volume of space. Both foci are spatially relevant, one in time zone A and the other in time zone B. This creates the appearance of elliptical orbits to describe the effects of gravity. Why only one focus is actually measured by humans will be eaten at the appropriate time. We can spit, at this point of the recipe, humans only measure in one temporal dimension.

Figure 12: Two time zones creating apparent elliptical space-time orbits. Each time zone is idealized as gravitationally equivalent circles.

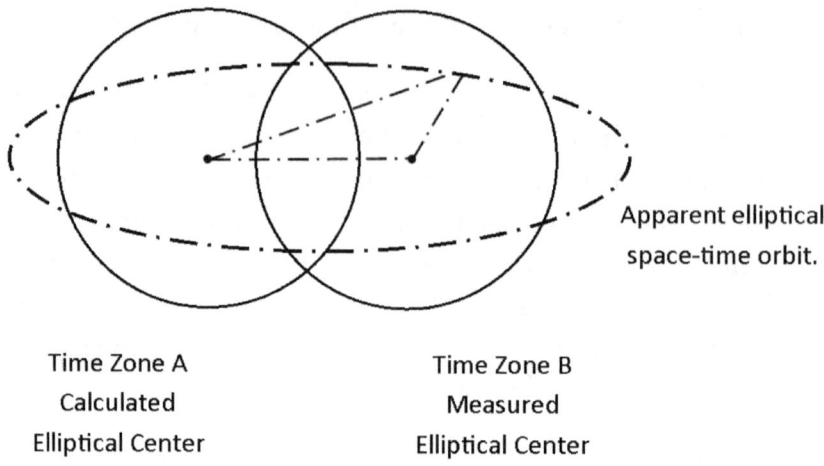

Apparent elliptical
space-time orbit.

Time Zone A	Time Zone B
Calculated	Measured
Elliptical Center	Elliptical Center

If gravity transforms either one of the idealized circles in its respective time zone, then the ellipse appears to retrograde through time.

Big-Bang
Reserve for Bundal Log Identification

If we were to fast-forward our geometric history we now know that a single point embeds a generalized Ψ-function that can represent an entire universe. It should be obvious this evaluation of a single, spatial place might be a bit overstated. Of-course, if an entire universe could be embedded into a single point, it would not take too long for someone to hypothesize that our universe actually started as a single point. You might say generalized Ψ-functions were doomed to explode into the pages of human history.

> Humans tell many jokes. Many revolve around sex. A plethora of Bundal jokes about humans revolve around God. We are the only species in the universe that would replace God with a dot and feel we have accomplished something significant.

Figure 13: Human being shouting to a dot-universe.

Some scientists have improved on the dot by indicating it may be a smeared out dot.

> The warning to artists is clear. Do not ever smear a dot. It might blow up.

> Hawking knows how much information can be carried on the surface of a black hole. How much information can be carried on the surface of a dimensionless, potential point?

Zero Omega
Reserve for Bundal Log Identification

What we need is an alternative understanding of triangles on curved spaces. A model with two, separate, temporal dimensions will allow us an opportunity to compare balls to saddles. For many of us riding on a ball is the same thing as riding on a horse. We tend to want to fall off. Nonetheless, this is far more fun than sitting on Euclid's flat plane.

> Omega refers to the curvature of our universe. If Omega equals one, then we live in a flat universe. All interior angles of a triangle sum to one-hundred and eighty degrees and π is sine-fine. If Omega is greater than one, then we live in a spherical universe. If Omega is less than one, then we live in a hyperbolic universe.

Triangles are well behaved in either a spherical or hyperbolic universe, but π seems more like an artifact from the flat-floater's world.

We will resolve the spatial issue of Omega by asking the dual questions of when Omega is less than one and when Omega is greater than one. We will then compare these answers across two temporal dimensions for a final resolution of the chord.

There are two things that challenge our credibility with the Bundals.

- We do not believe in God.
- We do believe everything was created from a single point.

We have not told them some of us believe both things at the same time.

Two temporal dimensions might reduce the number of spatial dimensions needed to draw triangles.

Our super-symmetrical world requires an anti-Kaluza-Klein methodology. Kaluza-Klein insights skim along the surface of temporal definitions while favoring the math as having strictly spatial import. Anti-Kaluza-Klein methodology, otherwise known as Occam methodology, focuses on temporal definitions.

It will be seen that typology is much more about the shape of time than it is the shape of space. Just look at one holed objects and you will understand what Bundals mean. Do a donut and a teacup really share the same typology as human textbooks indicate? Of-course they do. But how can they be so different looking to humans? Their difference is a temporal manifestation of the same typology. This implies the shape of space, as geometrically recognized by humans, arises naturally from temporal considerations.

Bundals use the same nouns and verbs to describe both a teacup and a donut.

The Bundal Recipe for Zero Omega
Reserved for Log Identification

The Bundal recipe assumes two planar spatial dimensions and two temporal dimensions.

Each spatial dimension represents a particular species. Each temporal dimension represents a particular species.

Bundals wonder if humans will start adding multiple dimensions to both space and time once we understand how to separate the temporal dimensions. They fear our mathematical quest to fully describe reality will never settle for simple numbers. Occam will turn over in his grave.

Bundal joke:

Question: How many dimensions do humans need to fully describe reality? Answer: As many as it takes.

Bundal joke:

Two human atoms debate over the ethics of gaining or losing an electron shell. The Periodic Table hangs on the outcome of the joke.

The test of understanding another culture is if you can laugh at their jokes.

The human sciences are due for a paradigm change. Human should SHFTSS that space and time are insufficient to describe reality.

Kuhn is a popular Bundal noun and verb.

Non-specialists suspect something is amiss as so many are excluded from the Pythagorean secrets of a four dimensional hyperspace that hides Calabi-Yau manifolds. What is needed is a simpler bi-temporal, bi-spatial model.

For this song we only need to identify the two dimension of time as A and B. Numbers are not needed. It feels like we are living in time C but actually this perception is based on the fact we are spatially connected to two discrete time zones.

> Pretend we live in both time zones simultaneously. Water and rice have made mush. We are convinced of the mush but would not know it was water and rice, unless we had been the ones to mix them.

Now let us savor the flavor of space. Imagine a sphere without a volume. Pretend you live on the outside of the sphere. Now pretend you live on the inside of the sphere. Pretend both places are temporally separated. The outside is in time zone A. The inside is in time zone B. There is no space between the outside and the inside as the separation is defined by time, not space.

Time zone A and time zone B are neither contiguous nor continuous. Their separation is defined by space.

The temporal and spatial separations are due to gravity and light.

Pretend that gravity molds space and time. Pretend that light contains the information to instruct gravity how to mold space and time.

> This might explain why photons are eternal.

> This might explain light cones.

This might explain why light is a limit to space-time measurements.

Now pretend that living in space and time is like surfing gravity waves. Pretend the semi-circle below represents an idealized gravity wave in two-dimensions. The separation of time and space (See Figure 20.) are molded by this gravity wave.

You are now prepared to understand the mystery of Omega.

Figure 14: Space-time curvature.

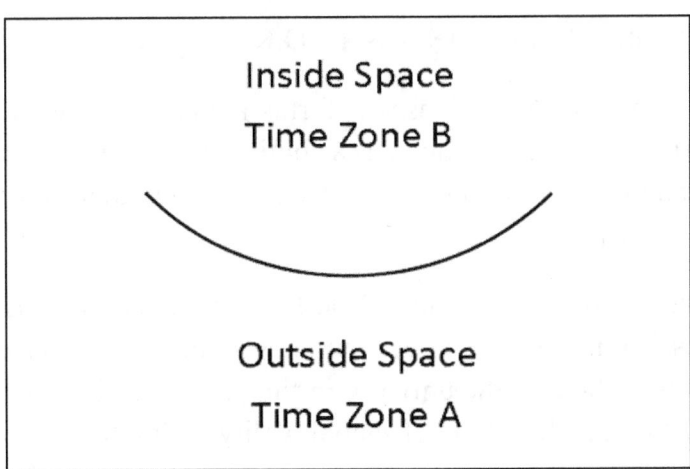

As bi-spatial creatures, we live on the inside of the wave with negative curvature. We also live on the outside of the wave with positive curvature. The outside pulls the inside towards it. To those who can only see the inside, space appears to be expanding into increasing entropy.

Bundals cannot explain it but they are quite convinced humans can only SHFTSS the inside. They seem incapable of SHFTSS-ing the outside. It is the outside that seems so dark to humans. Humans are luminous creatures. You would think they could at least see non-luminous reality.

From a Bundal vantage point the two time zones function as an out of equilibrium clock seeking rest. There is no rest on the inside. There is no rest on the outside. Light informs the gravity which forms the space-time measure.

- Tick: Time zone A condenses forming the inside space within time zone B.
- Tock: The inside then explodes to form a new time zone A. Welcome to the outside.
- See Figure 15.

God hates clocks. Calendars are O.K.

When time zone A condenses it forms a single indivisible whole within time zone B. Ironically, as a continuous concave surface, it is measured as a wave by humans since it is a continuous measure of one whole reality.

There are no particles in time zone B. It is one continuous whole. The classical notion of a Greek atom, as something indivisible, applies to the whole of the universe in time zone B. The psi-function is not a dot. It is the whole of spatial reality within time zone B.

Richard Feynman was correct concerning what parts of the universe a so-called particle might visit before selecting a slit. The so-called particle is physically everywhere in time zone B and only potentially somewhere in time zone A.

The measurement on time zone B is done from time zone A. The inside, as a solid if you please, is filled with potential, temporal energy, the source of what we know as the weak force, the strong force, and the electro-magnetic force. This egg shell explodes. The one piece of space matter becomes the one piece of temporal energy. A new time zone A is formed creating a new outside. The three, unified forces are actualized in time zone A.

This outside is filled with potential matter otherwise SHFTSS as actualized temporal energy. While humans would consider this to be a wave-like phenomenon, it is not continuous. Each place of potential has a probability to condense and locate itself in a temporally different spot once it condenses into a new time zone B. This non-continuous form of time zone A matter, with its potential-temporal separation, gives it a particulate phenomenology. The energy in time zone A will condensed and form a new inside creating a new time zone B.

Whitehead scholars might sing of concrescence rather than condensing. The concepts of pure potential, egression, and ingression are properly descriptive of temporal phenomenology as quantum mechanical in nature (Organism).

Matter appears as both a wave and a particle due to its existence within two, discrete, temporal dimensions.

$E = mc^2$ implies a relationship between energy and matter as due to a basic characteristic of light. The Bundals understand that characteristic to be the two primary dimensions, gravity and light, mold the two secondary dimension, time and space.

Matter appears as a particle in time zone A. Its particulate form is that of actualized temporal energy housed as potential space matter. Its appearance as a particle is an artifact of the human language used to describe it, e.g. mathematics.

Bohr's atom has been variously described as a point, a miniature planetary system, and a place of smeared out gases. All descriptions are that of a particle, although each description deviates from imagining a grain of sand. A point is dimensionless. A miniature planetary system seems to be an association of particles, rather than a single particular. And a place of smeared out gases hardly seems to qualify for describing a grain of sand. Bohr's atom shows the idea of a particle as a temporal manifestation is not new to human history.

Matter appears as a wave in time zone B. Its wave form is that of potential temporal energy housed as actualized space matter. Its appearance as a wave is an artifact of the human language used to describe it, e.g. mathematics.

There are not particles or waves in human mathematics. Mathematicians have yet to sign off on physicists claims to the contrary.

Substance is continuous in time zone B. There is no smallest particle within it, as shown by Planck.

The wave characteristic of a substance, particularly the idea that the whole universe is condense into a unitary substance, is due to the fluid dynamics of gravity. Potential temporal energy, as found in time zone B, has the sound of being played through fluid.

To Summarize:

The one time zone has temporal energy actualized. The other time zone has temporal energy potentialized.

The one time zone has spatial matter actualized. The other time zone has spatial matter potentialized.

The relationship between the two time zones is controlled by light through gravity.

$E = mc^2$ is an analog to the digital arrangement described above.

The initial ingredient for the above dish, time zone B, tastes similar to the ADM formalism for Relativity combined with Mach's Principle.

> There is always potential in space, even without matter. There is always potential in time, even without energy. There is always an actualization to both matter and energy. Julian Barbour's work is multi-choral, but still sounds the strain of no time in time zone B, per se.

Space and Time
Reserve for Bundal Log Identification

When you add a positive Omega to a negative Omega you come closer to a zero Omega.

The recipe for zero Omega comes together in these two half-dishes of time that always seem to function as exponents of each other.

> Time multiplies itself as space divides. Gravity adds or subtracts. Light remains a constant in mathematics. Operators are generally physically real. It is the numbers that create problems. One apple plus one apple plus one hammer equals one applesauce. The operators are real. The numbers mess up human minds.

> Whenever you see zero, there is always something.

Is a couch flat? Humans seem to think so since they sleep on them.

The couch lives in time zone B on a concave surface similar to the inside of a sphere. The interior sum of triangle angles adds up to less than one-hundred and eighty degrees. This feels cold.

The couch lives in time zone A on a convex surface similar to the outside of a sphere. The interior sum of triangle angels adds up to more than one-hundred and eighty degrees. This feels hot.

The couch is measured on the basis of the two time zones. The concave measure flattens out when temporally compared to the convex. The convex measure flattens out when temporally compared to the concave. Ideally, the final measure is zero, although slight temporal differences remain in actual measurement. The couch appears to be quite flat. The interior sum of triangle angles adds up to be one-hundred and eighty degrees. This feels neither cold nor hot.

Figure 15: The Omega Place.

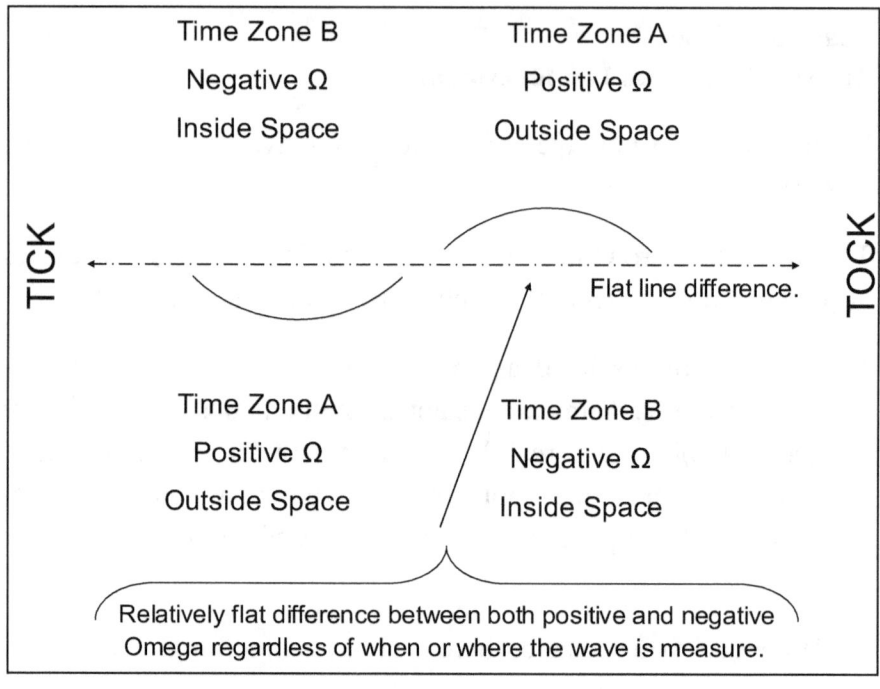

It is probable the outside is slightly larger than the inside. Omega may not be exactly zero.

This recipe SHFTSS an event as the temporal universe plus one. This accounts for a slightly larger outside than inside. Each inside grows when compared to the last inside due to the addition of events. As it explodes into a new outside, the outside has one more actualized event as part of its potential space matter. The outside is always larger than the inside.

The structure on the inside has an ever increasing tendency for low entropy in favor of focused life. The structure on the outside has an ever increasing tendency for higher entropy in favor of focused energy. As space and time grow, the place of low entropic life becomes ever unique.

By way of anticipation:

From the perspective of living within our drop of gravity fluid, it appears that space and time are growing.

From the perspective of Bundals, who can directly SHFTSS our drop of gravity fluid, it appears the gravity drop is evaporating into light.

The Four Basic Food Groups

Bundals SHFTSS four basic food groups. Their composition tastes like space, time, gravity, and light to them. Each food group offers two servings for a total of eight possibilities.

- Space offers an inside and outside.
- Time offers a past and future.
- Gravity offers a wet and dry.
- Light offers a hot and cold.

Bundals understand there to be four basic human food groups. These differ from the multitudes of categorization schemes used during the twentieth-century to define essential human food groups.

The human classifications system for life is far different than the Bundal understanding of a living universe. The twentieth-century has seen a multitude of categorical schemes used by humans for describing life.

Depending upon your country of origin life may be divided into six kingdoms: Animalia, Plantae, Fungi, Protista, Archaea, and Bacteria. Or it may be divided into five kingdoms: Animalia, Plantae, Fungi, Protoctista and Prokaryota, or Monera.

If you are a Phylogeneticist then none of the above is applicable. You care about monophyletic, paraphyletic, or polyphyletic taxa.

Biology is torn asunder by no less than Darwinian precepts.

Why do people turn to biology alone to try and figure out what kingdom they live within?

The twentieth-century has seen a multitude of categorical schemes used by humans for describing what humans should eat. Most humans possess bodies that will tell them what they need. Humans listen to other humans as to what we should eat and drink. Why?

We call it advertising.

Other humans do not possess our body. Bundals are confused as to why we would listen to them for what our bodies need?

Condescending to human concepts Bundals see four essential food groups for human life.

- Plants.
- Animals.
- Chemicals.
- Cannibalistic.

Humans can probably understand the plant and animal food groups. Chemicals are needed to provide what plants and animals do not provide. Sometimes the chemicals are added to the plants and animals before ingestion. Breathing oxygen or drinking water would count as a chemical addition. Cannibalism is used to describe processes where the body produces its own food. All four substances and processes are essential to human life.

The four essential Bundal ingredients necessary for universal life are space, time, gravity, and light. The universe is fluid. All four ingredients are found in bowls. Sometimes they are found on the insides of bowls. Sometimes they are found on the outside of bowls. The essence of life is drunk not chewed.

Humans have to chew and make liquid in order to swallow. Bundals can drink directly.

The composers opted for the Bundal rendition of using bowls rather than plates to eat. Bundals prefer to eat out of bowls. Some humans prefer to eat off of plates. Try using a straw to drink from a plate and you will better understand Bundals.

Ontologically there is:

- Inside and outside space.
- Past and future time.
- Wet and dry gravity.
- Hot and cold light.

All four food groups are needed to sustain life. When eaten life appears phenomenologically:

- Real.
- Present.
- Fluid.
- Warm.

The question of entropy deals with the bowls of light and gravity.

Phenomenology can become confusing. When all else fails we can fall back to thinking about what is certain about our place. Just ask René Descartes.

Ontology may be preserved for God Who is:

- Real, between Inside and Outside.
- Present, between Past and Future.
- Love producing. Gravity forms space and time to demonstrate love.
- Life giving. Light gives rise to life.

This music is contra David Hume. Paul Tillich may like it.

Ontology serves holographically for phenomenology.

Phenomenology serves iconographically for ontology.

The light bowl serves as a tuning fork. It is the fundamental tone. It does not always sound good to human ears. Perhaps that is why God's singing is interpreted by humans as speaking. The melodies are not complete without overtones. The first harmonic has no human harmony.

A creator requires a creation.

The gravity bowl serves as the second harmonic.

The time bowl serves as the third harmonic.

The space bowl serves as the fourth harmonic.

Thank God for the Pythagoreans.

You would think Edward Witten should win the Nobel Prize in music. How can music not bring peace?

Science is the activity of listening to the third and fourth harmonics as if they were one tuning fork. The sound is intriguing but deafens the senses. And the bowl seems full of holes. It is the first and second harmonic that enables the third and fourth.

Maybe now we have played enough to taste the possibility of being able to imagine a second temporal dimension.

Other Bundal Bowls

How do you answer questions of time?

Imagine Who, What, Where, Why and How.

Figure 16: Bundal question.

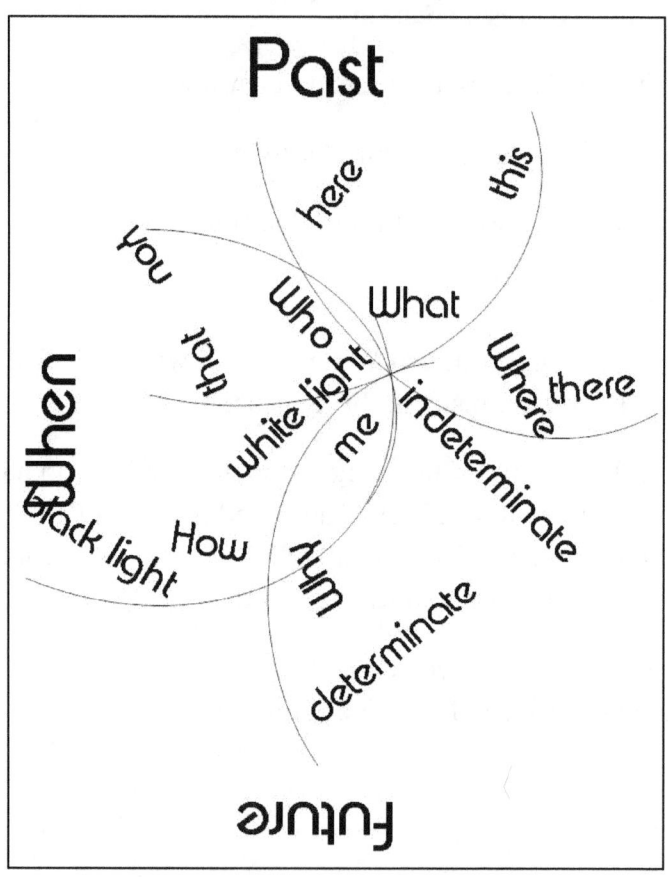

Figure 16 shows what composers are doing to decipher the Bundal SHFTSS. The Bundals have given this to us in forms we think of as words. Note the primary question is when. It is situated between the past and the future. The sub-questions are who, what, where, why, and how. Each of these sub-questions forms the centralized part of each bowl. Items within each sub-question are within each, particular bowl. The five bowls cross at only one point. This one point forms the answer to the question when, as framed by the past and the future. Bundal points have duration. The point identifying the event may be short or long.

Human speeches function similarly. One point may be short or long.

Human poetry might express the Bundal question in informal prose as:

Who is you and me.

What is this and that.

Where is here and there.

Why is determinate and indeterminate.

How is black light and white light.

The only thing missing is when.

When is past and future.

The answer to the main question converges as a result of all of the sub-questions. If the main question is on time, the answer must not use self-references to temporal events. Self-referencing is using a word to define itself. The time happened when says the same thing three times. There is no new knowledge or information. The temporal answer can only be ascertained by answering non-temporal questions.

Example:

When did the accident happen?

A proper answer to a temporal question involves:

The engineer was killed. (Who)

The train blew up. (What)

Morristown, Connecticut was mobilized to respond. (Where)

The diesel generator spun faster than its design ratings. (How)

Jacob overrode the safety accelerators on the train. (Why)

Answers to all of the above questions would tell a Bundal precisely when. Bundals do not use self-referential questions.

Humans would ask, "When did this happen? What did the clock say?" Such a question begs imprecision. Even humans know the Einstein-god plays cruel tricks. The clock on the train differs from the clock on the ground differs from the clocks of any witnesses. You will never know the when by referring to clocks.

When you answer a question of when with an answer of when your redundancy dooms you to imprecision.

- You are answering your question of when based on your SHFTSS of when. Human SHFTSS-ing is problematic. A temporal reading of a temporal SHFTSS misinterprets the SHFTTS.
- The proper use of SHFTSS to determine when follows. When is found once and only once at the intersection of who, what, where, why, and how.

Each question requires a holistic answer. When all of the answers resonate, a unique place in the past and future will be found without limiting the totality of the event between two artificial points in time. It is obvious a correct, precise measurement of time implies neither a short nor long duration.

This same process works for the other questions.

- Who is answered by what, where, why, how, and when.
- What is answered by who, where, why, how, and when.
- Where is answered by who, what, why, how, and when.
- Why is answered by who, what, where, how, and when.
- How is answered by who, what, where, why, and when.
- When has been sung already.

The main question always hangs outside of the bowls of the sub-questions. When all of the sub-bowls resonate, the main question is answered.

> This is the problem of human mathematics. Bundals see our numerical theories as necessarily self-referential. They prefer human narrative as narrative may be self-referential, but is not necessarily self-referential. Humans agree with Bundals concerning the self-referential aspect of mathematical theories. They call it the Gödel theorem.

Finally we are ready to imagine two temporal dimensions?

Sometimes it helps to think of fish rather than time. Every human who has ever fished knows the meaning of time.

Two Temporal Dimensions

The Bundals consider us to be a fish world. Humans are self-contained fishes.

We are fish of light but the cake must be cooked and cooled before we can ice it. We have tried to lay a foundation in the preceding chapters.

The Bundal experience of tasting two temporal dimensions is best experienced as a drink from the fountain of gravity.

The Einstein-god created rubberized rulers, variable-speed clocks, trains, and elevators to help us appraise three-thousand years of human understanding.

Does something new replace something old? Or does it merely transform it?

You can see rulers, clocks, trains, and elevators.

How do you see time?

Maybe our ability to imagine more than one, temporal dimension will propel us into a deeper yet simpler exploration of the reality in which we exist.

Two temporal dimensions do not dispense with what the Einstein-god created. In fact the concept places the notions of Einstein firmly within quantum mechanical behavior.

Whales have known how to communicate across our planet for millennia. Bundals can hear across the universe.

Their string ensemble will assemble to play through the clearest medium of all, a fully colored gravitational fluid. The entire universe listens.

Fish World
Reserve for Bundal Log Identification

Imagine an aquarium. The water in the aquarium has dual characteristics categorized under one term, temperature. A god-like view of the aquarium proves the aquarium to be hot and cold. The hot side is hotter than hell. The cold side is slightly negative of zero degrees Kelvin. The fish might say that one side of their universe was absolutely hot and the other side was absolutely cold.

The fish did have a religion where the gods controlled the aquarium. Boreas was the god of fate, always freezing things into proper position. Vulcan was the god of possibilities, always heating up the need for decision making. The more Boreas blew cold the hotter Vulcan's fires froze. The fish were always caught between fate and decision making. This was all based on the temperature of the aquarium.

The temperature could go up or down, but the direction always came from Boreas. All fish knew temperature flowed in one direction, from Boreas to Vulcan.

The fish scientists did not believe in fish gods. The fish scientists had equations that proved temperature could flow in two directions. It seemed as if Boreas and Vulcan had equivalent buoyancy. But the common fish remained convinced you could never really travel back into fate once you had made a decision.

Science fishing was great entertainment. SF novels and movies depicted fish actually visiting Boreas or Vulcan and changing the temperature line of their world.

The aquarium is hot and cold. The fish are hot and cold. Fish thermometers are hot and cold. The atoms of their universe are hot and cold. Every fish cell is hot and cold. Every environmental factor is hot and cold.

Hot and cold could even bubble up from nothing.

So what is different about this aquarium?

The fish gods never got along. Neither would share what the other had. There was no warmth. Only hot and cold existed.

There is a perpetual divide between the cold sides and the hot sides. The cold is not heated. The hot is not cooled.

The fish thought they lived in the warmth. They actually lived in hot and cold between two frightening gods.

As the gods fight, the aquarium and the fish experience warmth in varying degrees. Warmth is the accidental creation of two gods who will never get along. Considering this as a perpetual fight, it is as if warmth spontaneously arises from pugilistic chaos.

When a fish swims in the aquarium, no matter what direction the fish swims, there is always a hot and cold side within the fish, on the fish, and outside of the fish. This demarcation of dimensional temperature is not directly evident to the fish.

It is difficult for any fish to recognize the environment within which they swim.

Each temperature functions as its own species of a dimension. The two species of temperature are discontinuous which is a fancy way of saying temperature is discrete. Ontologically the temperature is discrete as either cold or hot. Phenomenologically the temperature feels like warmth.

Fish cosmologists debate whether the temperature of the aquarium is getting hotter or colder.

Light refractions through the water indicate the fish barely feel ninety-five percent of the temperature available within the aquarium.

Some fish hypothesize there is a wetness ether through which the light refracts. Wet fish in a wet aquarium cannot measure wetness. The ether remains undetectable to them.

Temperature can be measured by using either a cold scale or a hot scale developed by fish scientists.

Early cave fish learned how to make things warmer. This made them feel like masters of the aquarium.

Amazingly, if the fish decide to measure the cold they can get a reading on the cold scale. If they decide to measure the hot, they can get a reading on the hot scale. But they cannot get a simultaneous reading on both the hot and cold scale. It is as if warmth is dependent on measurement by either scale before it can exist.

The above mystery is further complicated by the fact that two fish, in two different locations, might read a different temperature.

Worse, temperature readings seem to jump. They are never continuous. Every fish knows temperature is continuous, so why does it jump?

All attempts to unite these phenomena into a unified fish-theory have proven elusive.

There is a promising new technology in fish world that offers to create a total fish theory. It is the theory of drops. It believes the fundamental components of the aquarium are drops. Unfortunately, more and more types of drops are being hypothesized to model the aquarium. Pretty soon there will be more drops named than there is space in the aquarium.

Religious fish are chafing over fish scientists. They accuse the fish scientists of starting their own religions. Fish scientists have brought this on themselves. They can create a multitude of drop models while believing each model must exist somewhere at some time. Some even talk of multiple aquariums existing in parallel to their own.

It does not help that the fish scientists continue to search for the god-drop.

The multiple aquariums are connected by warm-holes.

The fundamental constituents of matter are corks. Their floating properties give rise to all the buoyancy within the aquarium. Somehow buoyancy and temperature are connected.

A Feynman fish pioneered the idea of networking. This fish likes to think of hot and cold as functioning as independent sources of information. Fish live at the interstices. The idea of a network as a physical reality does not appeal to most fish.

Fish philosophers are always arguing whether there is one fundamental aquarium or two fundamental aspects of the one aquarium.

Our fish live in two temperatures. But they can only sense one that can apparently travel in two directions. One direction is warmer. The other direction is cooler.

Imagining Two Temporal Dimensions
Reserve for Bundal Log Identification

Perhaps time's two apparent directions tell us what lies at the ends of the same line. The past and the future do not form a continuous line. They are separated by space. They form two, distinct temporal dimensions. There is no present.

Another way of saying this is that our space exists in the past. Our space exists in the future. Our space does not exist in the present.

Our sensory organs make sense of the past and future as one present.

- Two eyes see one picture.
- Two ears hear one sound.
- Two fingers feel one object.
- Two nostrils smell one odor.
- Two sides of the tongue taste one flavor.
- Two hemispheres of the brain SHFTSS one reality.
- All of the organs have multiple-sensory inputs.
- All of the organs have a mono-reality output.

Science, religion and ethics are deeply affected by SHFTTS.

- Science is the attempt to find the monadic interior based upon the mono-reality output. This is a partial lie. Partial lying is endemic to science. The laboratory must exclude the rest of the universe to be relevant. The notion of a closed system admits the lie.
- Religion is the attempt to experience the total exterior based upon the multiple-sensory inputs. This is a partial truth. The giving of partial truth is endemic to religion. The full truth is we do not know much about God.
- Ethics is the attempt to create. History is the story of ethical experiments.

Imagine swimming in a universe where every atom, every molecule, and every cell within and without your body lived in the past and the future. No matter which way you swam or which direction the universe swirled around you, both are always in the past and the future. Neither is ever in the present.

The present is a projection or sensation of two, underlying, bifurcated, temporal dimensions. It does not exist. Only the past and the future exist.

What would it be like to swim in such a universe?

It is not necessary to conclude *a priori* or *posteriori* that the space of the past functions like the space of the future. Both spaces would have contextual differences. As each space interacts, each brings a quality from the bowl of the temporal time zone they inhabit.

- A hypothetical creature may have a head in the past and a tail in the future.
- Or visa-versa.
- A hypothetical creature may have a front in the past and a back in the future.
- Or visa-versa.
- A hypothetical creature may have a top in the past and a bottom in the future.
- Or visa-versa.

For the sake of argument, our creature has an inside in the past and an outside in the future.

This forms a bi-spatial, bi-temporal creature.

Bundal joke: Quadrupeds are stable while bipeds balance between worlds. Quadrupeds have two feet in each dimension.

Such a creature could remember the past and predict the future. They would be unable to hold the present. Every time they would try to grasp the present, or anticipate its arrival from the future, they would end up looking at empty hands.

Every time they would SHFTSS the present they would only have remembrances of the SHFTSS. They would also have anticipations about the next SHFTSS. The difference between the anticipation and the remembrance is called disappointment or exhilaration. Emotions are a functional way to differentiate between the past and future.

Such a creature would argue whether its existence is determinate or indeterminate. As a creature of the past many arguments could be marshaled in defense of determination. As a creature of the future many arguments could be marshaled in defense of indetermination.

Such a creature would argue whether God plays with dice. As a creature of the past the game board appears fixed. As a creature of the future the creature wonders if there even is a game board.

Such a creature might notice there are apparent jumps in space as the unwitting difference between the two temporal time zones.

Such a creature might believe that the past extends backwards while the future extends forward. The creature would fail to realize the two, temporal dimensions point towards each other, rather than in opposite directions.

Figure 17: Human vs. Bundal time lines.

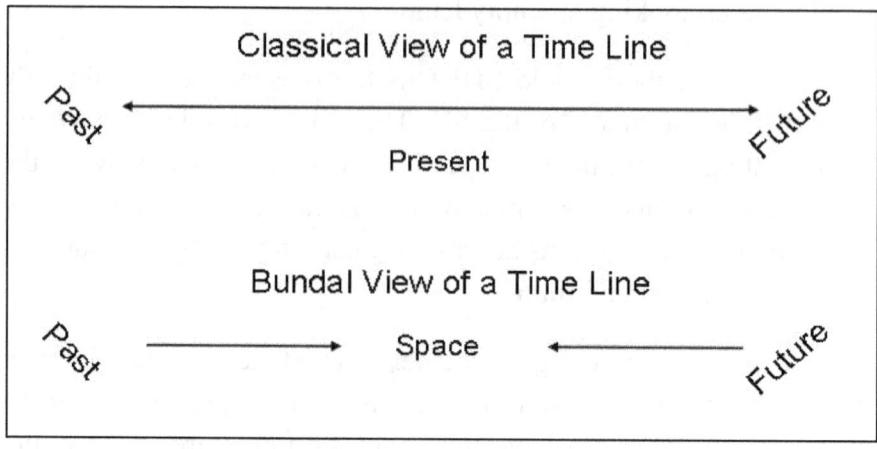

SHFTSS the song of two dimensions and you will taste answers to many hungers.

Almost all philosophical questions, historically raised by humans, are answered correctly. Some of the answers are applicable as a result of living in the past temporal zone. Other answers are applicable as a result of living in the future temporal zone. Demanding that one set of answers is correct while the other set of answers is incorrect is to deny a bi-temporal, human existence.

Quantum Mechanics Made Foundational
Reserve for Bundal Log Identification

Our creature wants to add a quantum mechanical foundation to relate space to the concept of two, temporal dimensions.

We can see rulers, clocks, elevators, and trains. We cannot see time. We will SHFTSS our picture of two-separate temporal dimensions with insights rather than sight. The final picture is but a Bundal-inspired model.

Let's take the Einstein-god to his inner limit and freely interchange space and time.

> Some people insist the Einstein god is female. They blame his wife.

We will pretend that in the past space has consolidated into a single whole. Another way of saying this is that the universe functions, in the past, as the classical Greek view of an atom. Space is indivisible. It is as if there is no time in the past. It has all been absorbed into forming a fully determinate space. Time is expressed as a potential energy held within the space it inhabits.

> For sake of argument we will call this space the inside space.

Space explodes into a new future. Space becomes a temporal field largely structured by the form of the last space. In a way there is not space in this explosion, only future time and the potential for space. In this future time zone decisions are made that will influence the forming of the next past. Since the decisions are not formalized until the creation of the next, new past, the future is characterized as in-determinate.

> For sake of argument we will call this space the outside space.

- The future seems to have no solid space. It is all a temporal energy field housing potential space.
- The past seems to have no time. It is all solid space housing potential time.
- Points of space are phenomenological, not ontological.
- Instants of time are phenomenological, not ontological.

Future time collapses and forms the next, new past, filled with a single space.

One could muse there is a Planck difference between the past and the future.

How long each particular dimension retains its peculiar formation is unknown to us. You cannot measure time by time. You cannot measure space by space.

Those living in Sèvres, France are living the partial lie of science.

Bundal Joke:

Questions: What galaxy do you have to live in before a meter in Sèvres, France becomes problematic?

Answer: Your own.

This process might be interpreted as a recurring big-bang theory taking the idea that the universe functions like a clock to a new level.

The tick is that the universe condenses into space.

The tock is that the universe explodes into time.

We live in the tick and tock but it seems like we live in the present, the hyphen between the tick and tock.

The temporal measure of our universe is one past deep. The past is not ontologically continuous since its space explodes to form the next future time. The past ceases to be except as a memory, preserved in the field structures of the next future.

Heidegger scholars would face the hyphen as Dasein. The past is present-at-hand. The future is ready-to-hand.

There are not two or more pasts connected to each other. The one past is only connected to the one future. The past and future are connected through space.

The temporal measure of our universe is one future long. The future is not ontologically continuous since its space condenses to form the next past time. The future ceases to be except as a measured event within a solid space, housed in a potential for the next future. A new future is created when the new past time explodes.

Whitehead scholars would feel the condensation as the concrescence of the organism.

God has two temporal balls to play with. One is fully determinate and known to us as the past. The other is full of dice and known to us as the future. The past shapes the next new future just as the future forms the next new past. But the balls are separate on an ever changing game board. The rules are few and simple.

- Condense
- Explode

Time is an amusement park which takes space for a ride. The park is fenced in by light and the shapes of space-time are formed by gravity.

Does time run forwards and/or backwards? The answer is no.

Space-time equations ignore temporal discreteness. The mathematical transformation of space-time allows forward movement to account for the explosion of the past into a new future. The mathematical transformation of space-time allows backward movement to account for the condensation of the future into a new past. Since temporal discreteness is disregarded, every past looks continuous and every future projects as continuous. The principle of sameness is axiomatic enabling the equations to work in similar conditions. Whether this redundancy is fortunate or unfortunate depends upon whether you are a K-Meson.

The future and past are always primordial.

103

Memory and prediction are functional.

Measuring What?
Reserve for Bundal Log Identification

What is it that humans measure?

Bundals would think that humans could measure both the past and the future. Bundals would expect that humans could measure both the inside and the outside. Bundals were surprised this was not the case.

For some reason the human- relationship to the future-outside is that of

- Sight-less
- Sound-less
- Feel-less
- Gustation-less
- Smell-less
- Sense-less

Bundals regard us as fish of light that can directly sense the past-inside of our aquarium. We swim in the past-inside and the future-outside. Due to a strange form of evolution, we tend to navigate through the future based solely on the past. We are incapable of directly apprehending the future-outside.

> The Bundals explanation for our strange form of evolution is sung elsewhere.

> Our SHFTSS is directly related to the past-inside. We remember the past and measure the inside.

> Our SHFTSS is indirectly related to the future-outside. We predict the future and infer the outside.

We SHFTSS from the past and can perform surgery on our brains.

We SHFTSS from the future. We are incapable of deciphering consciousness as a future reality. Some of us believe it emerges from the past. We experience consciousness but are unable to determine its source. It comes from the future.

Regarding the past, we can measure one past against another past. We do our measuring from the future. The future embeds the quality of memory into the past. Quality measurements are made possible because both past events are fully determined in remembered relation to each other. Since both past events form dual reference points, time can seem to flow bi-directionally between the two measures. This is not that time flows in either direction between two past events. It is only that our memories allow a measurable temporal relationship between the two. And our measuring devices depend upon the perspective of the memory we choose to use.

Different people will have different memories due to differing perspectives.

The legal profession has known this for centuries.

A video of an egg falling off a table and breaking can be run backwards to make it appear the egg reassembles itself on the floor and jumps back up to the top of the table. This is a realistic measure between two past, never to be replicated, events as recorded on film. One past event had a whole egg. Another past event had a broken egg. If every time we ran the video, either forward or backwards, it kept changing on us, then we would have grave doubts about how fully determinate either recorded event was.

There was one remembered-measured past when the egg was whole. There was another remembered-measured past when the egg was broken. The so-called temporal distance between them remains the same no matter which memory is given temporally-remembered priority.

We can be assured of our measurements because they measure one past to another. Measuring the quality of one past to that of another past is simply to ask the difference between two fully determined experiences. It is obvious measuring two fully determinate realities, even while disregarding the rest of the universe, ought to have some level of built-in accuracy.

Human measurements comparing two or more pasts are said to be accurate to within plus or minus some margin of error. The measurement errors may be due to perspective. They may be due to memory. They may be due to intrinsic redundancy since one past space is not continuous with a prior past space. Nor is one past time continuous with a prior past time.

Our lack of errors in measuring may be due to miracles.

We must not confuse measurement with prediction. We measure the past. We predict the future. The future is predicted on the basis of probabilistic studies. The probabilistic studies are based on comparing multiple pasts.

Probability works best when space is large and time is short. The rate of change in probability, as space shrinks and time lengthens, is a constant.

What does it mean to bra-ket the past and future? The bra-ket itself needs a context beyond space and time.

Anger

"We are not insane. We just do not know."

This ejaculation took place in the context of a misunderstanding of one composer over an interpretation of a Bundal art piece.

Some of us are beginning to hate Bundals.

Human Ants

The Bundals fear what we will do to the universe with temporal engineering.

The Bundals fear what we will do to ourselves when we discover how disabled we are as a species.

Figure 18: Home of the finest species to ever evolve.

Limited Creatures

Reserve for Bundal Log Identification

Humans can only see the inside.

This means that whether we use a microscope or a telescope or a horoscope, we are viewing the inside.

- We are seeing our future through a microscope: Multiple mille-seconds into our future.

108

- We are seeing our future through a telescope: Multiple millenniums into our future.
- We are seeing our future through a horoscope: Multiple machinations into our future.
- The third dimension for humans is a temporal projection.
- If the farthest star starts to fall, the Einstein-god will make sure it takes more than 13.7 billion years to hit us.
- Those stars are our future. They are not our past except as a unified whole.
- Medicine knows those cells seen through a microscope are our future.

Humans say we cannot see the future.

The future is about all we do see.

When the rock is heading towards your head, you are seeing the future.

We are seeing a temporally projected future from within the inside past. It is a future filled with many decision points. It is never the future that will be finally chosen. You may duck and the rock will miss your head. Maybe.

We are not seeing the true future within which we live, and move, and have are being. That future can only be seen from the outside. While we live in the future-outside, we do not see the future-outside.

Bundals are continuing to struggle with the aftermath that humans actually exist.

Math seems unique to humans. Math is the human attempt to see the future.

The Ant Colony
Reserve for Bundal Log Identification

All ants can only SHFTSS half of a glass.

We use to be the center of the universe.

Then we were the center of our solar system.

Then we marched around the sun that gave us life.

> Our solar system is in an arm of our galaxy. An arm! Not the heart. Not the head.

> It seems as if our galaxy might be in a rather obscured part of the universe. In fact everything seems obscured to us.

Even our universe may be a crumb of food in gravity and light compared to all of the other universes that may exist.

To believe that ants perceive less than half of their place in the universe is ego-crushing.

How humiliating if their perception is limited to five percent of the universe.

Humans are disabled SHFTSS-ers.

Positivism and animism are humbled.

> Bundal Joke: Ants have antennas. To be reincarnated as an ant would help humans see more of the earth.

> If dirt tastes bad then why did God give ants antennas?

Crushed Ants

If we are unable to SHFTSS all aspects of our participation in space and time, just imagine how insignificant we will feel when we discover that space and time are themselves inadequate to create reality.

> Is the apple of God's eye a small black spot on an all-white canvas? Should not the all-white canvas be congratulated on creating a small black spot? Is not the small black spot special?

Crushing the Ant:

- Inside-Outside to Inside-Only. Measures only the Inside.
- Is peripheral even within the Inside.
- Spatially adept to create stuff that disintegrates.
- Recognizes that space and time are insufficient for understanding reality.
- Humans treat their physics with gravitas. What is gravity?

Gravity can certainly crush an ant.

Einstein Worshipped

The Einstein-god enabled us

- to see rulers that shrink and elongate.
- to hear clocks that tick slower or faster.
- to feel elevators that fall down or push up.
- to experience the ride on sub-luminal trains.
- to spice the flat taste of Euclid.

The Einstein-god did not play with dice. Apparently the Einstein-god was a control freak.

The Einstein-god did a Hegel maneuver to save Parmenides and Heraclitus from choking.

- The never changing universe was spit out of Parmenides.
- The ever changing universe was spit out of Heraclitus.
- The ingredients were mixed to form invariance. Everything changes so as to stay the same.
- The spirit of Hegel lives on.

Parallel Universes

The composers feel the need for an environment to contextualize relativity within quantum mechanics.

Are there two parallel universes?

The answer is indeterminate based upon what ingredients are necessary to bake a full-fudged (fledged) universe.

Humans can be entertained with the following fiction.

Fiction is another song that sounds exactly like model. The taste of fiction is indistinguishable from the taste of a model.

Bundals wonder about the integrity of human's SHFTSS-ing towards direct apprehension of the inside. Models feel like nonfiction to many humans, especially those who are mentally advanced. This contagion is being spread throughout the human race, by the mentally advanced, through viruses called avatars.

The issues we struggle with throughout our dilemmas can now be played again. This time, the chords have resolution.

Are there parallel universes?

The answer is indeterminate. The answer depends on the taste of a full-fudged universe.

Imagine the inside-past as universe B.

Imagine the outside-future as universe A.

The inside is connected to the outside through time.

The past is connected to the future through space.

113

Space-time is not ontologically unitary.

Space-time as connected to space-time is somewhat abrogated. Space is connected to space, only through time. Time is connected to time, only through space. You should not confuse the two in your imagination.

The questions of when and where are answered separately. They cannot be both answered at the same time and in the same space.

To imagine that space functions in an inverted or converted relationship to time is to forget the place of gravity and light.

Each universe has its own equations. Each universe may be connected through operators or constants.

A simple matrix of simple matrixes appears quite complex until the simple notions are clearly established. Having the correct operators between matrixes should always produce unique results. Having the same results proves only a non-existing, static universe. Having similar results does prove we can only measure in one direction, from the future to the past.

To expect to find the exact same thing in each universe is to beg the necessity of two universes.

To expect to find the exact same event in each universe is to beg the necessity of two universes.

To expect to find space and time in each universe allows the concept of universe as opposed to the concept of a bi-verse.

Euclid's parallel postulate has always been imaginative. Riemann proved the notion of parallel can mean anything.

Bundals do not have a SHFTSS for parallel.

The Bundal's four dishes of space, time, gravity, and light are nutritious only at their intersection.

Universe B has a planar form of matter that is projected through time to appear three-dimensional. The Einstein god should be well pleased.

Universe A has field loops that are projected into space to appear three-dimensional. The Einstein god may be less pleased with this.

1 Universe A + 1 Universe B = 1 Universe

EPR rang the bell for non-local realism.

Universe A has three dimensions, two spatial and one temporal.

Universe B has three dimensions, two temporal and one spatial.

1 Universe A + 1 Universe B = I Universe of 4 Dimensions

The Bundals say arrrrrghhh. The Bundals say arrrrrghhh not because the ideas presented here are wrong. They say arrrrrghhh whenever humans use numbers to describe reality.

Another way of saying this is there are two, parallel dimensions of gravity and light that form two, parallel dimensions of space and time. These parallel dimensions form our universe, best seen by Bundals as but one drop of gravity fluid.

Temporal Projection

Temporal projection is easy to imagine.

A Phenomenological Imagination

Truth One: If space were three dimensional, there ought to be vast voids within the three dimensions.

Truth Two: Substances should clump close together leaving vast voids throughout space.

If you looked in any one direction only, there ought to be some directions that have nothing to be seen.

If space were three dimensional, there ought to be no voids within the three dimensions.

Substances should diffuse throughout space.

If you looked in all directions, every direction would look the same.

Both of the above truths are true. Is this redundant?

> Background radiation seems the same no matter what direction you look.

> There appears to be vast voids between galaxies.

Space must be three dimensional.

An Ontological Reality Check

If space is floating in time, then one can expect either clumping and/or diffusion. Clumping and diffusion cannot both be ontological realities unless there was a God particle interfering with the whole situation.

116

If there is a God particle, then diffusion surrounds us from before the time the God particle was created. Clumping surrounds us once God got involved.

Clumping in three spatial dimensions implies large areas of void. Humans ought to be able to look in at least one direction, through a straw, and see no objects. Why?

There is no such thing in the universe as a completely solid, three dimension object. Not even a three dimension atom is solid.

> We ought to be able to live on a nucleus and look in at least one direction and not see our electron moons.

Clumping should be so uneven there should be some areas, viewed in a single, narrow line of sight, where no spatial objects would be seen. You could still see the diffused background from the time before God intervened but you ought to be able to find areas where space is not clumped.

Imagine Projection
Reserve for Bundal Log Identification

The one-hundred percent lack of solid three dimensional substances indicates some non-spatial material, if it can be called that, separates space. The non-spatial material could form the appearance of an additional third spatial dimension assuming the existence of only two spatial dimensions. If space were solid in either one or two dimensions, then we would expect to find objects in every direction within those one or two dimensions. Once temporally projected, those objects appear to clump together in every direction.

Clumping does occur as seen through the Hubble Deep Field Collection. Everywhere we look, even through a straw, numerous objects can be found. If everything was clumping within three spatial dimensions, we would expect to find some views with no objects with our view frame. There ought to be some lines through the universe with nothing on them.

If space is two dimensional, and the third dimension (depth) is temporally projected, then the clumping in two spatial dimensions implies we will see something (eventually) along every straight line we look.

Temporal projection can be thought of as a series of parallel movie screens, each one showing us a more distant (as in time) potential future. The result is:

- The clumping of objects is seen everywhere in every direction.
- The clumps occur because they appear to be separated by the non-spatial material.
- The temperature of all spatial material will differ, but the temperature will be felt in all directions.

Figure 19: The Hubble Telescope Deep Field Collection as the narrowest straw in the universe. Only a few billion galaxies are seen through a straw.

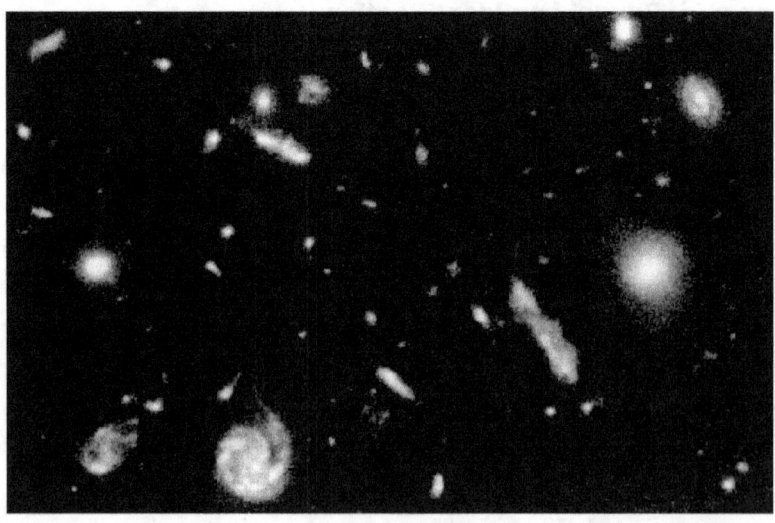

Imagine a sliced loaf of bread. Pretend you are sitting on the inside heel of one end of the bread. It is a two dimensional surface you are sitting on and are a part of.

Pretend you can see through all the slices of the loaf but you are not sure you can see the other end. You do not really see the demarcation for each slice as much as you see a continuous and deep look through the loaf. Each slice represents a potential future as compared to your slice of bread. These slices are coming to you. They will each in turn condense to form a new end slice, upon which you may still sit.

The farthest slices will take the longest to reach you. They will also carry the potential for the most total change, before they reach you. Each slice of bread is but a temporal projection of the space you sitting upon. Space does not have depth. It only appears to have depth.

Credits

No Reserve for Bundal Log Identification

The Hubble picture was uploaded on human time July 26, 2012 from http://www.stsci.edu/ftp/science/hdf/hdf.html. Material credited to STScI on this site was created, authored, and/or prepared for NASA under Contract NAS5-26555. Unless otherwise specifically stated, no claim to copyright is being asserted by STScI and it may be freely used as in the public domain in accordance with NASA's contract. However, it is requested that in any subsequent use of this work NASA and STScI be given appropriate acknowledgement. The composers, Author, and Bundals credit NASA and STScI.

Spinning Matter

In the mathematics of chromodynamics spin is a unit-less number. The author posits it is unit-less in space and time but not in gravity and light.

The Bundals will not tell. They do not like numbers.

Amazingly even classical rotation may be an artifact of gravity and light rather than mass. This concept would offer an alternative answer to the dilemma of Newton's bucket at Mach speed

This composition is particularly difficult. Many of the notions here are explained elsewhere.

The inside is matter.

The outside is anti-matter.

As temporally distinct, their electromagnetic signatures are reversed. Their spin characteristics depend upon gravity and light. Why? Space and time are derived from gravity and light. Gravity and light function to create space and time.

Matter and anti-matter only annihilate each other when temporally touching.

The unitary inside space may have some flecks of anti-matter. This is an admission that the process of condensation is an idealized SHFTSS that does have real world anomalies.

The unitary outside time may have some flecks of matter. This is an admission that the process of explosion is an idealized SHFTSS that does have real world anomalies.

Spin is not spin. Physicists know this.

Spin may be whole or half.

Inside space and time are orthogonal.

Outside space and time are orthogonal.

If inside space explodes into outside time and inside time explodes into outside space, then the relationship of outside time to outside space is orthogonal.

If outside space condenses into inside time and outside time condenses into inside space, then the relationship of inside time to inside space is orthogonal.

With so many orthogonal relationships, it is easy to come up with whole numbers and half-numbers.

The explosion and condensation process mimics rotation.

Rotation is a gravitational illusion.

Rotation is an artifact of unitary space.

Rotation is an artifact of temporal differentiation.

Newton's spinning bucket of water begs the question: Rotation around what?

As an artifact rotation can appear inside or outside.

Rotation implies the next past will be different.

Rotation implies the next future is just around the corner.

Humans come up with really funny numbers whenever they go around in circles.

Spin is a measurement that points to a reality beyond space and time. The illuminating darkness of light and the expanding compression of gravity molds space and time.

Spin measures differences occurring external to the space dimension and time dimension. The spin difference does not reflect internal space and time processes. Spin is related to internal space and time processes as they are molded by light and gravity.

Humans can only measure an actual difference between two pasts. Spin is a spatial-temporal memory of the activities of gravity and light.

The intrusion of a measuring device is insufficient to describe the difference associated with spin.

When spin is maintained as a set of numbers within space and time, it must remain unit-less.

Unknown to the fish, there are processes occurring on their aquarium that mold the temperature and buoyancy within their aquarium. Such processes can be felt or measured, particularly as it comes to the reversal of hot temperature and cold temperature. Since these processes occur within a context beyond temperature and buoyancy the numbers remain non-defined except for mathematical language seeking reality.

Fish are thankful spin does not exist within their universe.

If a neutron star, with a circumference of thirty-thousand meters, can spin at nine-hundred thousand revolutions per second just imagine how fast our universe must be spinning. Both thoughts raise skepticism concerning the taste of classically described rotation.

Bundals SHFTSS humans go around and around in circles as they try to catch the next event or measure the next particle.

Why apologize for spin not being spin in quantum field theory? Rotation may not be rotation in classical physics. This will lead to a different SHFTSS of gravity. Only a mono-temporal creature believes gravity arises out of space and time. Newton's vortexes are an indication of gravitational influence on space and time.

Time and Space as Nouns

The next few chapters represent the meat of Tubal's recipe.

What do words mean?

Gravity is attractive.

Gravitas is heavy.

Is a graviton really two-thousand pounds?

> Humans found Robert Heinlein in *Strangers in a Strange Land* intriguing. Bundals found it boring.

> Who does not grok? Humans call it cannibalism. Bundals call it recycling.

Explanations of reality abound throughout human history. Quality explanations require the creation of new words. Bundals consider most human words superfluous. The composers have tried to honor the tastes of Bundals and use words common to late twentieth-century, American, English reading humans.

Use of these words represent such a small portion of the human race, the composers see their music as a niche scratching an itch. The use of portmanteau is irritating. So do not scratch.

> Physics is filled with made-up words.

> Does a wavicle flutter? Is a wavicle flatter?

> A wavicle is not a wave. A wavicle is not a particle.

> Why would anyone swimming through this composition complain about the use of phenomenology to bridge ontology?

125

General relativity compares acceleration to gravity. If you accelerate in an opposite direction, should not this produce anti-gravity?

We live in a two-dimensional spatial world, temporally projected to give us our third dimension of SHFTSS. Any launch, in any direction, from our two-dimensional launching pad, yields us acceleration felt as gravity. We are accelerating into our future, compressing future space by expanding past space, to bring our future place closer, faster. Having two spatial dimensions, instead of three, makes the process acceleration the same as for gravity. The future is coming at us sooner. Future space is being compressed.

The composition moves on to wordiness, a solo without music. The Bundals remain concerned so many words are being used in this recipe to describe physics. At least they are funny words, at least to Bundals.

Enfolding Dimensions

There are two orders of dimensions.

Arrrrrghhh (pronounced arg′-gŭh) is a technical word. It is used to describe the Bundal's disdain for using mathematical terminology. There is no equal to the word within the Bundal vocabulary.

A Bundal, if they could, would say there are two sets (arrrrrghhh) of dimensions that are enfolded.

Cantor beware.

The typology of the enfolding is unknown to humans. The typology can be conceptualized as conversion.

- Past time will convert to outside space.
- Future time will convert to inside space.
- Outside space will convert to past time.
- Inside space will convert to future time.

126

Conversion leads to super-symmetry without a simple dualism.

Why do humans call it super-symmetry if the particles are not symmetrical with sparticles?

Figure 20: Enfolding of time and space.

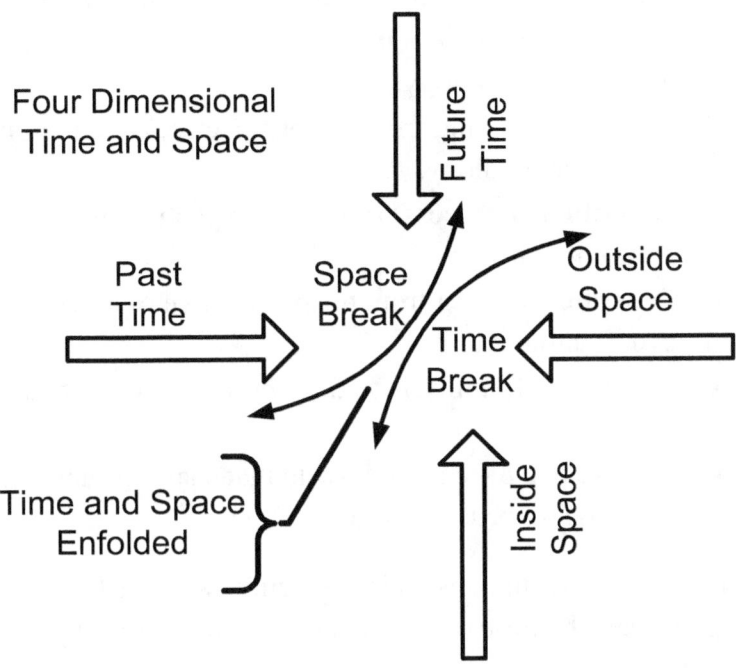

Enfolding can be conceived as the juxtaposition of the breaks. As described later, there are two juxtapositions with three M-dimensions each.

Bundals use nouns and verbs.

- The conversion of dimensions will be described as nouns.
- The conversion of dimensions will be described as verbs.
- Conversion implies correlation.
- Enfolding is described as the mixing of nouns and verbs.
- Describing a process from the perspective of a noun implies a substance ontology.
- Describing a thing from the perspective of a noun implies a substance ontology.
- Describing a process from the perspective of a verb implies an event ontology.
- Describing a thing from the perspective of a verb implies an event ontology.
- The story of reality is told by bringing nouns and verbs together.
- We live in a bi-temporal world that shares a substance-based and process-based ontology.

Philosophers, theologians, science-fiction writers, literary authors, and poets have historically created new realities by simply changing nouns into verbs and verbs into nouns. Is it a wave? Or is it a particle?

Knowing made-up words can earn students high grades on academic exams. Unfortunately, when you have to make up words, you are forcing foreign realities upon your readers.

Made up words are not as insightful as words commonly used by humanity.

Nouns Enfold Verbs
Difficult to Place as Part of the Log Identification

There are two different types of harmonies. We will discuss the second-ordered harmonies (arrrrrghhh) most used in three-thousand years of human understanding.

> The Bundals wanted the composers to dispense with the word, "discuss." Humans use discussions to count the number of angels on the head of a pin. Discussing is disgusting to Bundals. The Bundal *teleos* aims at effective communication. Words seldom accomplish this feat, especially among humans.

> It hurts the Bundals for the composers to use the word "recognizable" instead of SHFTSS.

> The composers will cease to use arrrrrghhh unless necessary for clear translation. Please know the Bundals continue to shiver whenever numbers are used.

The second-ordered harmony is space and time.

The first-ordered harmony is gravity and light.

> When gravity and light and space and time cancel out their ontological collisions, what is left is the phenomenon humans understand as sensory. This will be described towards the end of the recipe.

Second Ordered Dimensional Nouns
Bundal Log in Pain

What immediately follows is a presentation of time and space as nouns. Time zones are on the left. Space zones are on the right. This will remain consistent if you hold it up to a mirror.

Figure 21: Time and space as nouns.

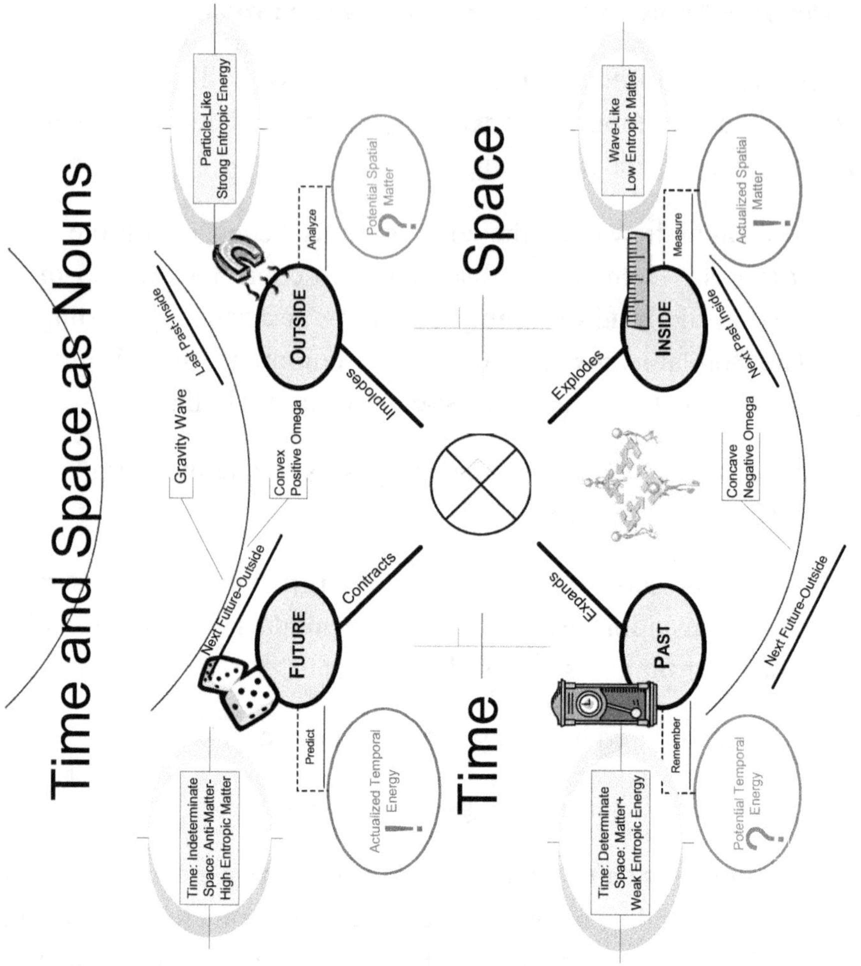

Space is composed of two-enveloped species.

- The one species is the outside.
- The other species is the inside.
- The spatial species are temporally distinct.

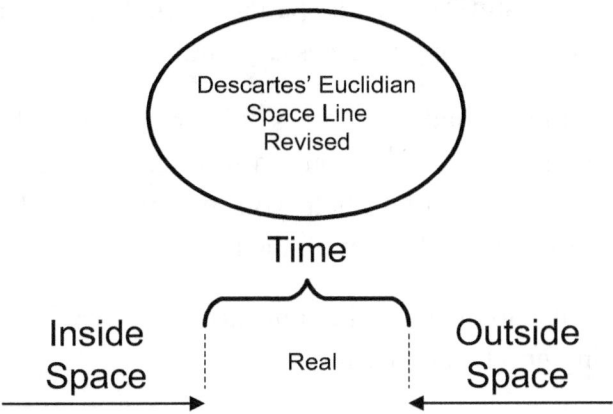

Figure 22: Descartes' space line revised.

Time is composed of two-enveloped species.

- The one species is the future.
- The other species is the past.
- The temporal species are spatially distinct.

Figure 23: Newton's time line revised.

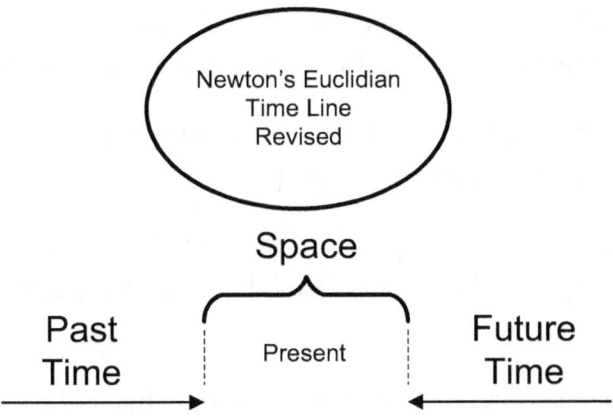

When the four species of space and time come together, they create a real present. Notice that the real refers to time and the present refers to space. Their enfolding (inter-action) creates the experience (phenomenon) of a real present. Twentieth-century physics misunderstands the present to be a temporal concept and misunderstands the real ("stuff") to be a spatial concept. This is a category mistake that creates many unnecessary ontological paradoxes.

Humans might wonder if there is only an outside and inside then is the real real? If there is only a past and a future then is the present present? It might help you to remembers the Einstein-god always had problems with the now.

By way of anticipation, this question will be addressed when discussing gravity and light.

The future and the outside enfold each other. This is referred to as the future-outside.

The past and the inside enfold each other. This is referred to as the past-inside.

The reason the past and inside uniquely enfold is due to the verb forms of space and time and is discussed elsewhere.

The reason the future and outside uniquely enfold is due to the verb forms of space and time and is discussed elsewhere.

The following is an attempt to relate the dimensions Bundals use to classical, twentieth-century human physics.

The future-outside is composed of three mathematical dimensions. One of them is spatial and two of them are temporal. These are called M-dimensions.

The past-inside is composed of three mathematical dimensions. One of them is temporal and two of them are spatial. These are called M-dimensions.

Anytime you see the expression M-dimensions or M-dimensional, know you have discovered a mathematical description for a dimension. This contrasts sharply with the Bundal propensity for valid, ontological descriptions of dimensional reality.

Space is a Bundal dimension. The M-dimensions and their derivatives are x, y, and z.

Time is a Bundal dimension. The M-dimensions and their derivatives are x, y, and t.

M-dimensions tend to change depending on whether they are in Greek or Latin superscript and/or subscript. The final determination of M-dimensional thinking should be resolved by theologians who excel in combining Greek and Latin insights.

The three M-dimensional structure of each enfolding occurs as a result of:

- Two M-dimensional future time converts into two M-dimensional inside space.
- One M-dimensional outside space converts into one M-dimensional past time.
- Two M-dimensional inside space converts into two M-dimensional future time.
- One M-dimensional past time converts into one M-dimensional outside space.
- The future-outside equals three M-dimensions.
- The past-inside equals three M-dimensions.
- Future time contracts to become a new inside space. (See Figure 21.)
- Inside space explodes to become the next future time. (See Figure 21.)
- Past time expands to become the next outside. (See Figure 21.)
- Outside space implodes to become the last past time. (See Figure 21.)

Space becomes time and time becomes space. This is the meaning of conversion.

The four terms, contracts, explodes, expands, and implodes, refer to processes of space and time that have their context within gravity. These four terms guarantee that conversion does not imply invariance or mere equality. Conversion implies transformation whose processes may be invariant, but whose final results are unique. This is discussed elsewhere.

> The future-outside enfoldment can be envisioned as riding the convex side of a gravity wave. This would lead to a positive omega for the universe. (See Figure 15.)

> The past-inside enfoldment can be envisioned as riding the concave side of a gravity wave. This would lead to a negative omega for the universe. (See Figure 15.)

> The intersection of a real present, remembering the Einstein-god, is relatively flat.

Outside Space

The outside can be conceived as a particle like line. This is quite Euclidean in thought although the points are real. It is life in one M-dimensional space.

An argument for an understanding of real particles has many holes. Dividing matter into increasingly smaller portions begs the question of infinity. While modern Euclidians would approve, Planck's reality is not nearly so kind. It is as if Euclid's reality describes points in terms of potential which is why there is always some point between any two. But Planck's reality states there is a mystery between any two real points. This mystery reveals the fact of a process between discontinuous space and discontinuous time.

Particles are normally conceived as forming real matter. This implies, but does not prove, that particles are, in and of themselves, real matter. It is a tautology to say that matter is made from matter. So it is no surprise to discover particles are converted into time. The Einstein-god would be proud.

The outside, particle-like line is best conceived as potential spatial matter. It is difficult to discern parts and wholes. It does not take a complete potential to produce an actual whole. It only takes selected parts of a potential to produce an actual whole.

Bundals grudgingly admit human science has stumbled upon the concept of actual wholes being produced by selected portions of any potential.

Another example is more pernicious:

Human science states it this way: You can disregard the rest of the universe as insignificant in favor or what goes on within an isolated experiment. Actual results can be obtained by paying attention to only a few, selected phenomenon.

Shoot a bunch of potential spatial matter, otherwise known as particles, through a double-slit, and they actually make a whole spectrum. The spectrum is well defined, as opposed to haphazard, both by temporal considerations and the contextualization of space and time within gravity.

Humans tend to analyze outside space using statistical data. It resides in future time. We SHFTSS outside space through the temporal projection process. We are SHFTSS-ing our future potential.

Inside space cannot SHFTSS inside space. It is a unitary whole.

Outside space cannot SHFTSS outside space. It is a particulate potential.

Outside space as potential matter has strong entropic energy. Line length does not determine the quantity of points. The concept of entropic energy and entropic matter is discussed elsewhere as a relationship between pressure and volume due to gravity and light. Suffice it to say potential matter implodes to form a one M-dimensional past time.

Potential spatial matter converts to become potential temporal energy.

Ethical Perspective

As an aside to the present discussion:

Plato's real forms are jettison is favor of real spatial and temporal potentials. Each real potential is attached to and makes possible particular, Aristotelian, instantiated substances and events.

Outside space, as a real potential, is dimensionally integrated to instantiated future time.

Past time, as a real potential is dimensionally integrated to instantiated inside space.

The mystery of Plato's cave and shadows in found in the light, not the cave nor the shadows.

Plato and Aristotle were both correct.

Past Time

The past can be conceived as a fully determinate temporal dimension. In a sense it is locked into place.

It is one M-dimensional time.

It is formed from the imploding of one M-dimensional outside space.

136

As fully determined it possesses weak entropic energy.

Humans tend to remember the past.

When the outside converts to the past, it forms a next past.

When the past converts to the outside, it forms a new outside.

Figure 24: Conversions of outside space and past time.

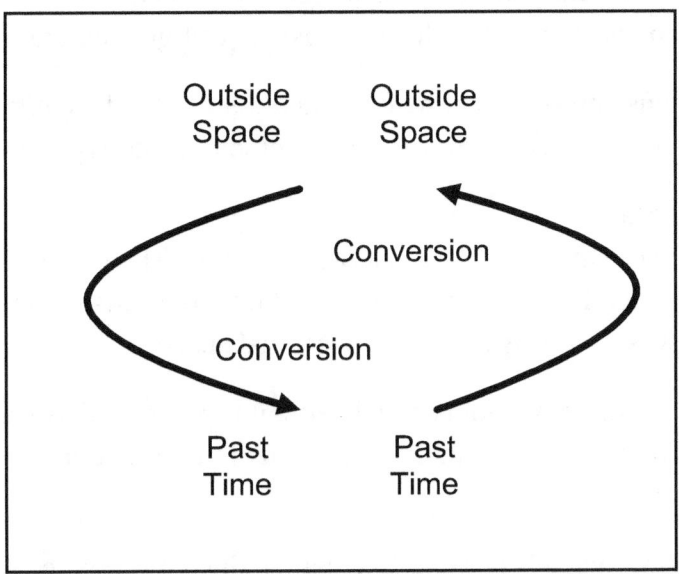

Outside space is one M-dimensional. Past time is one M-dimensional. Humans can sense in the conversion of any dimensional species an artifact known as rotation.

Each next past is not contiguous to the next future. Nor is each last past contiguous to the past before it. The past and the future are connected spatially. The most recent past and the past before it are spatially remembered.

Each new outside is not contiguous to the new inside. The outside and inside are connected temporally.

This disjunctive, ontological relationship between space and time is Whitehead's notion of nexus combined with Heidegger's *gelassenheit*. The composers have attempted to avoid philosophical language, such as Whitehead's concrescence, in favor of macro-chemical language, such as contracts. The use of macro-chemical language obscures the quantum mechanical foundation that so deeply informs the time and space relationship. Our composition, functioning as an imaginary model, will demonstrate dynamics is directly relevant to issues of light and gravity.

This discussion of past time is limited due to the difficulty we humans have to conceive of temporal geometries and typologies.

Inside Space:
Inside space can be conceived as a solid plane. The plane is not necessarily smooth or Euclidian. It is a solid whole. It is unitary. There are no parts to it. It is two M-dimensional space.

As a solid whole, without spatial parts, it is prehended as wave-like. Wave-like matter is spatially continuous in two M-dimensions and temporally set.

Outside space is particle-like matter since it is single M-dimensional and temporally separated.

This addresses particle-wave duality in the context of two, temporal dimensions.

The inside is best conceived as actualized spatial matter.

The inside-past has the quality of low entropic matter.

Its changes can be measured from the current past time to another remembered, past time. Changes are measured from the future.

The Past-Inside is a unitary whole. The inside offer two M-dimensions. The past offers one M-dimension. For those paying

138

attention, this totals the three, M-dimensions of twentieth-century physics. Time is potential energy and determinate. The unitary whole of space is locked into the current, unchanging past time until the next, past time is created.

Future Time:
Future time is converted from inside-space.

Inside space is converted from future-time.

Both are two M-dimensional. The orthogonal lines in figure 25 are meant to remind the viewer if the two, M-dimensional aspect of inside space and future time.

Figure 25: Conversions of inside space and future time.

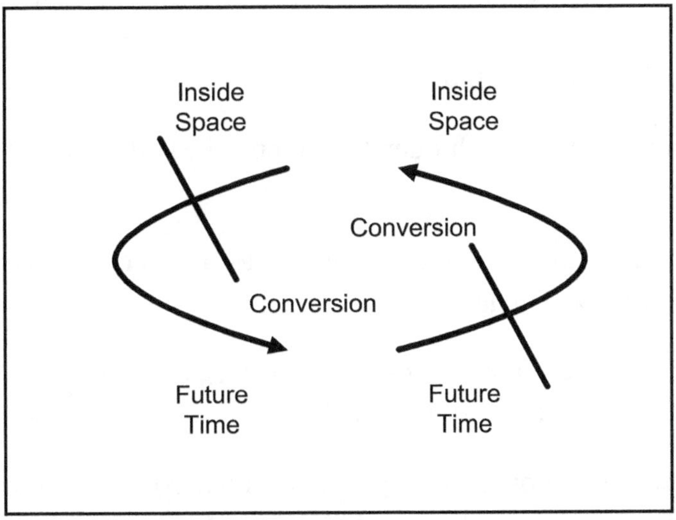

Future time forms a field around outside space. Future time offers two M-dimensions. Outside space offers one M-dimension.

Future time is actualized temporal energy. Temporal energy is the fundamental energy of the three unified forces. It is the energy that enables change to occur from one inside space to the next inside space.

Future time contracts to form inside space.

Inside space explodes to form future time.

There are two forms of matter. Each form is noted as having opposite charges and opposite quantum spin. The opposite charge is an artifact of light. The opposite quantum spin is an artifact of gravity. Light and gravity will be discussed later.

As a matter of convenience the Bundal model considers inside space to be matter.

As a matter of convenience the Bundal model considers outside space to be anti-matter.

The difference between matter and anti-matter is a matter of temporal description rather than a spatial designation. Matter is past time. Anti-matter is future time.

Anti-matter is typically hidden to humans due to their ability to only see the past-inside.

Anti-matter can exist in the past-inside as an exception to the general rule that permits less than the ideal.

Matter can exist in the future-outside as an exception to the general rule that permits less than the ideal.

Processes of creation are always messy and leave behind continuous reside that may inhabit temporal and spatial regions contra-productive to its continued existence within those regions.

Matter normally exists in the past because the past is determinate. Any anti-matter in the past is a residual product and may not be present in the next past.

Anti-matter normally exists in the future because the future is indeterminate and spatial matter is potential. Any matter in the

140

future is a residual product and may not be present in the next future.

As creatures of light we are made of matter and anti-matter. We do not annihilate as long as our inside and outside stay temporally separated.

As creatures of light we are made of matter and anti-matter. We do not annihilate as long as our past and future stay spatially separated.

The particular drop of gravity fluid we know as our universe keeps space and time in contra- phase with time and space.

Visualizing Second Ordered Dimensions
Reserve for Bundal Log Identification

The cross bars in figure 25 are meant to remind the reader that inside space and future time are two dimensional.

One of the composers sees three-M dimensional representation in terms of a DNA helix. Inside space and future time represent two, spiraling ribbons connected by two bars, outside space and past time. It must be remembered that neither the ribbons nor bars are ontologically continuous except for conversion. They are phenomenologically continuous as three classical dimensions in time.

Light and Gravity as Nouns

Bundal Log in Pain

First ordered dimensions are fundamental to second-ordered dimensions.

The first ordered dimensions are light and gravity. Light carries the information telling gravity how to form. There is a feedback loop not described in the Bundal model.

Space and time are second-ordered dimensions. They are derived from gravity and light.

> The composers recognize any model that states that light informs gravity is of the highest order of fiction and vain imagination. That unicorns exist offers less speculation. The composers are unable to make complete sense of the Bundal model. We do seek to remain consistent with our purpose of SHFTSS any model that would allow humans the ability to see two temporal dimensions.

The two species of light and the two species of gravity work together to form space and time.

If there are other universes with other laws of physics, it is because light and gravity have worked together to form other types of universes. Space and time are derivative of gravity fluid. Space and time are not the only derivatives of gravity fluid. The information necessary to form other types of universes is carried by light.

> Since the information necessary to form other universes composed of nouns and verbs other than space and time is not part of our own universe, the ingredients of such universes are not able to be imagined by humans. Imagination

is based on information. Imagination does not create information. Imagination lies about information.

The speed of light is a limit to energy and matter because it is a first ordered dimension.

Gravity does not unify with the three unified forces because it is a first ordered dimension.

C. Auguste Dupin, in *The Purloined Letter*, would say the anomalies of light and gravity indicate the obvious location of the letter. It is out in the open and quite visible as something of a different quality than mere space and time.

This is a presentation of light and gravity as nouns.

A presentation of light and gravity as verbs will follow.

Figure 26: Light and gravity as nouns.

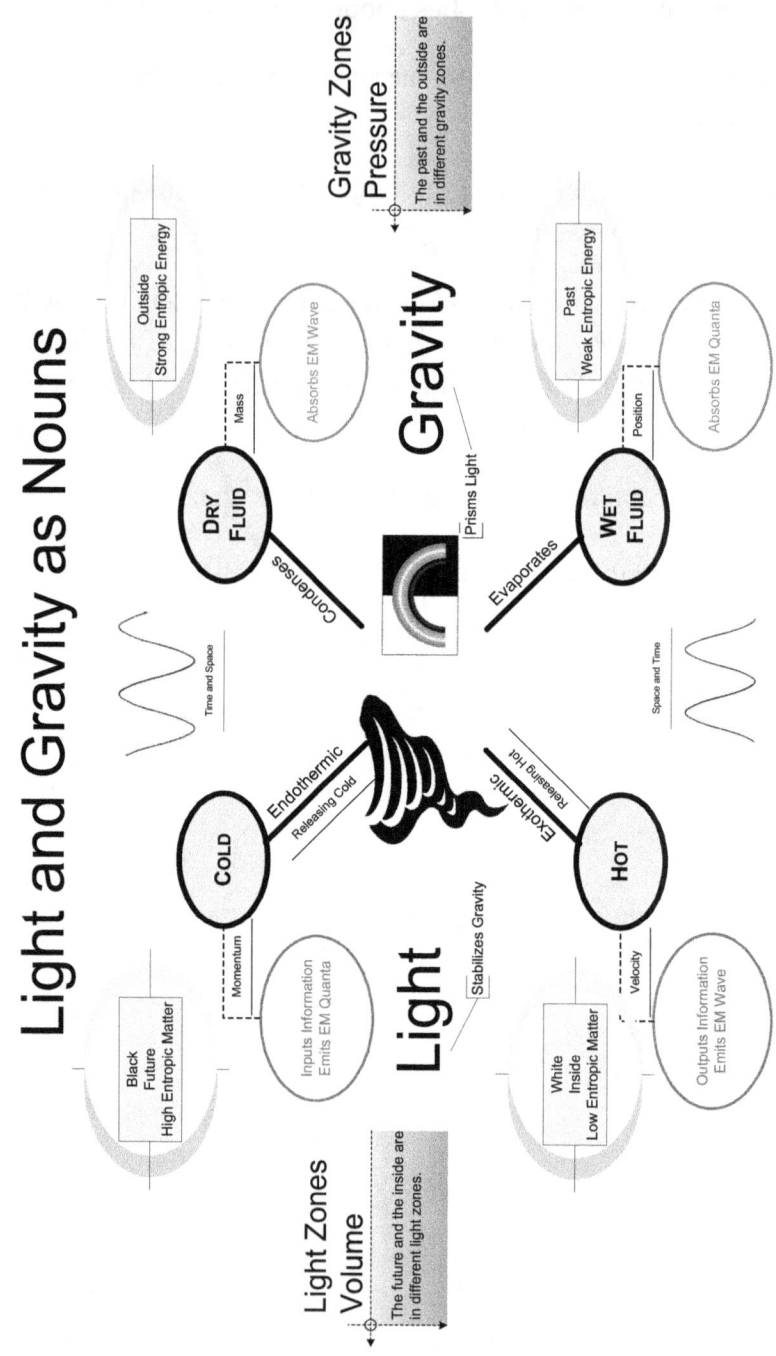

Gravity is composed of two species.

- The one species is wet.
- The other species is dry.
- The two gravitational species are distinct as separated by hot and cold light.
- Gravity is a fluid. It is the fundamental force that causes space and time to have the qualities of a BEC, solid, liquid, gas, and plasma.

Light is composed of two species.

- The one species is hot.
- The other species is cold.
- The two light species are gravitationally distinct.
- Light carries the creative information of the universe.

Entropy is a measure of energy and matter.

- The outside has strong-entropic energy.
- The past has weak-entropic energy.
- The inside has low-entropic matter.
- The future has high-entropic matter.

As a wet fluid, gravity evaporates creating cold light. As a dry fluid, gravity condenses creating hot light.

It is acceptable to think of cold light as black light.

It is acceptable to think of hot light as white light.

Black light and white light combine to create the phenomenon of color, or EM radiation.

The black future is evidenced by high entropic matter and is SHFTTS-ed as emitting EM quanta.

The white inside is evidenced by low entropic matter and is SHFTSS-ed as emitting EM waves.

Black cold light serves to input information.

White hot light serves to output information.

Black cold light and white hot light are not continuous. They are separated by gravity.

The light zones are characterized by volume.

Future time and inside space are in different light zones. Their light zone context gives rise to their qualities as temporal and spatial nouns.

Cold light is endothermic. The process for cold light is not that of releasing heat into its environment so as to cool down. The process is that of releasing cold. As the process is future, humans do not directly apprehend or know of cold as a quality that can be released. We misinterpret it as absorption of heat away from an object.

> Both isothermal and adiabatic processes are possible depending upon perspective. The nineteenth century caloric theory needs revisited.

The cold serves as an input to information.

Cold is released into the past. Humans are capable of knowing there is a limit to cold in their phenomenology. Humans call it absolute zero. They do not question why there appears to be no limit to hot. Hot is released into the outside. Humans do not directly apprehend the outside.

Light stabilizes gravity. Gravity prisms light.

As an input of information, cold has direction. It offers momentum.

As an output of information, hot has velocity.

Gravity as a dry fluid absorbs EM waves from hot white light. As hot light releases hot, mass is produced as strong entropic energy on the outside.

Gravity as a wet fluid absorbs EM quanta. As cold light releases cold, position is attained as weak entropic energy.

The past and the outside are in different gravity zones. Their gravitational context gives rise to their qualities as temporal and spatial nouns.

The interplay of two species of light and two species of gravity give rise to the context for the past, future, inside, and outside.

The three unified forces of space and time are different expressions, due to context, of the one temporal force. The one temporal force is derived from gravity.

The particle families of space and time are different expressions, due to context, of the one spatial matter. The one spatial matter is derived from light.

> Eight planets. Eight electrons. We are always looking at the inside.

The wave-like and particle-like attributes of matter are directly derived from their place in two different time zones.

The wave-like and particle-like attributes of matter are directly relational through the interaction of dry fluid gravity and hot white light.

The mathematical notions of commensurability and incommensurability are SHFTTS as due to perspective. The Bundal model requires more than a shift of perspective. It requires an acknowledgment of the ontology of different time zones as the basis for apparent shifts of perspective. The basis of different time zones finds its context within gravity and light.

Unit measure has always been problematic in the mathematical sciences. Poor Pythagoras. Pick any unit. It is incommensurable with something, somewhere, at sometime.

The wave-like and particle-like attributes of matter are indirectly derived through the interaction of wet fluid gravity and cold black light.

The Bundal model dares to differentiate between entropic matter and entropic energy. This is consistent with the mysteries and equations of Boltzmann and Prigogine.

The composers are uncertain whether luminous is a quality of both white and black light. The composers are uncertain whether non-luminous is a quality of both white and black light. If luminous applies only to white or black then it may serve as a replacement noun. If non-luminous applies only to white or black then it may serve as a replacement noun. If both luminous and non-luminous apply to both white and black light, then they serve as adjectives. The adjectives would be luminous and non-luminous.

The composers understand luminous to refer to the type of matter and energy that humans can see. The composers understand non-luminous to refer to the type of matter and energy that humans cannot see. Twentieth-century human cosmology sees a ratio of nineteen parts non-luminous to one part luminous in our universe. Humans are luminous. This ratio gives rise to a deep suspicion that the difference between black and white light, coupled with an ontological difference between cold light and hot light, that twentieth-century humans are misunderstanding the difference between luminous and non-luminous matter and energy.

Verbs Enfold Nouns
Difficult to Place as part of the Log Identification

It is one thing to describe the four basic food groups as nouns. Any such recipe is inadequate to discerning the quality of the ingredients. Verbs must be used to make complete sentences.

The four basic food groups will now be described as verbs. Combining nouns, and verbs, with adjectives and gerunds will make complete sentences. Portions of the story line of reality can be completely and uniquely described with the four dimensions.

Gravity and light form first ordered dimensional verbs. Space and time form second ordered dimensional verbs.

There is no priority implied between nouns and verbs. Substance ontologies and event ontologies are both needed to make complete sentences within the ontological discipline.

East and west are not wrong. Both are correct.

Art and science are not wrong. Both are correct.

Every human culture perceives truth. Why we kill each other over it remains a mystery as a grave lack of appreciation towards our fellow human beings.

Light and Gravity as Verbs

This is a presentation of light and gravity as verbs. Verbs related to light are on the left. Verbs related to gravity are on the right.

Figure 27: Light and gravity as verbs.

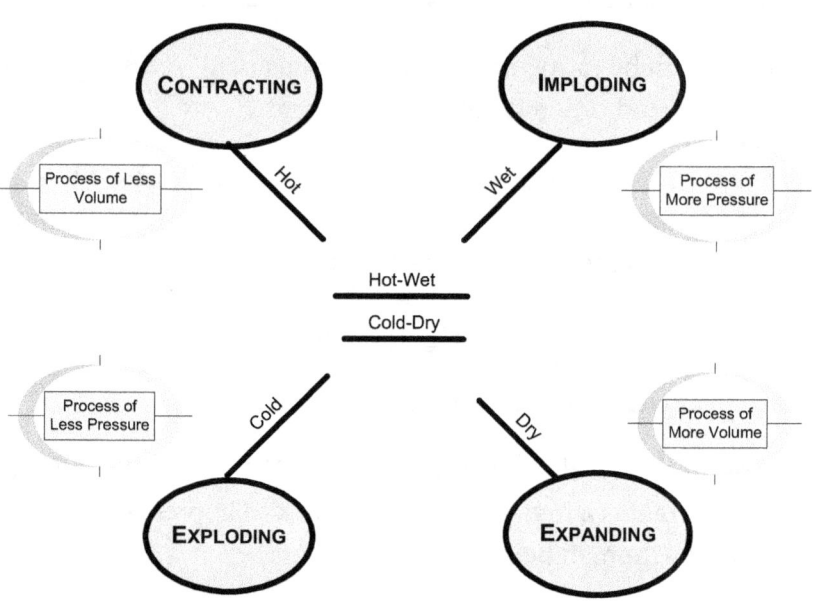

First Ordered Dimensional Verbs

Bundal Log in Pain

Light has two processes. There is contracting light which results in less volume. There is exploding light which results in less pressure. The less volume makes things hot. The less pressure makes things cold.

Gravity has two processes. There is imploding gravity which results in a process leading towards more pressure. There is expanding

150

gravity which results in a process leading towards more volume. The process of more pressure makes things wet. The process of more volume makes things dry.

Think of the difference as being between de Sitter space and anti-de Setter space. Time might be reversed within the Bundal model should de Sitter space be compared to anti-de Setter space.

The interaction between contracting light and imploding gravity creates the past-inside. Hot-wet relates to the past-inside

The interaction between exploding light and expanding gravity creates the future-outside. Cold-dry relates to the future-outside.

The process of light on gravity produces an apparent rotation.

Masses that move can always be thought of as rotating, however imprecisely, around any supposed point of reference.

The mystery of gravity remains in the Bundal model. The composers are unable to decipher Bundal songs. As a noun, gravity may be associated with a second set of verbs. Those verbs are known as whirl and swirl. As a gerund we might say gravity is whirling or swirling.

The composers are uncertain whether whirl is a quality of both wet and dry gravity. The composers are uncertain whether swirl is a quality of both wet and dry gravity. If whirl applies only to wet or dry then it may serve as a replacement verb. If swirl applies only to wet or dry then it may serve as a replacement verb. If both whirl and swirl apply to both wet and dry gravity, then they serve as gerunds. The gerunds are whirling and swirling.

Newton's bucket of water is the result of the way gravity works between its two species. Humans can only directly SHFTTS the past-inside. Of-course it looks as if the rest of the universe, the

151

outside, was causing the water to continue spinning after the bucket comes to relative rest to the observer.

Figure 28: Gravitational whirl and swirl.

Detail: Gravity as a Gerund

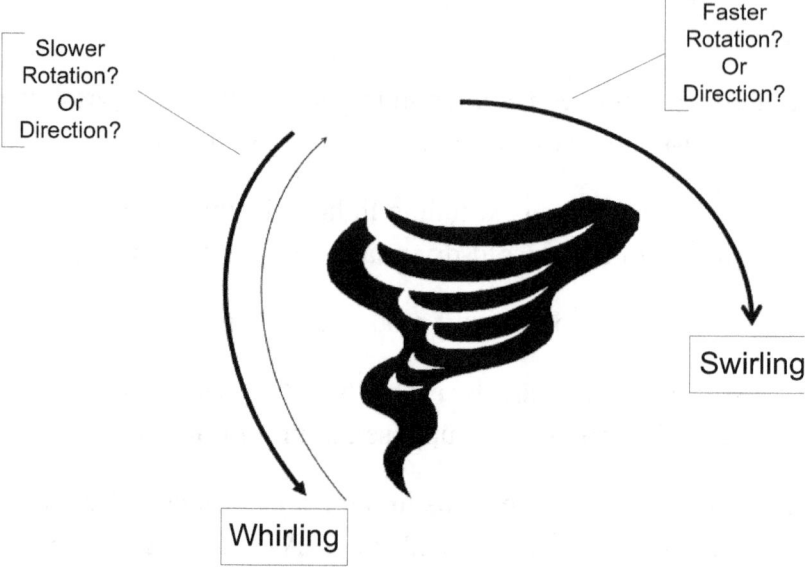

Whether whirling and swirling are ontological rotations seems unlikely. Whirling and swirling may just be directions. They are directions only in the sense they may be orthogonal resulting in half-spins, so-to-speak.

If whirling and swirling are opposites, then half-spins are an artifact of SHFTSS through time and space.

The fact of gravity SHFTSS as imploding and expanding indicates opposition of internal direction. When indirectly observed through time and space, gravity appears rotational.

Time and Space as Verbs

Figure 29: Time and space as verbs.

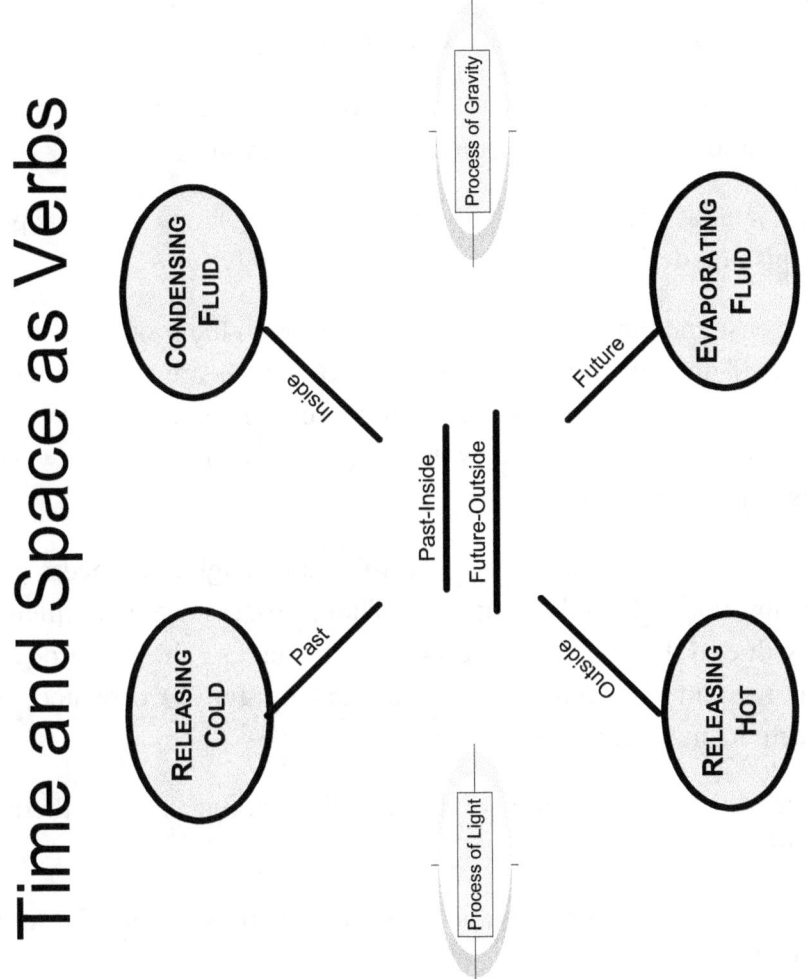

Time and space are second ordered dimensions. They are derived from light and gravity.

Time has two processes as a verb. Time can be seen as the process of releasing cold to create a new past. Time can be seen as the process of releasing hot to create a new outside.

Space has two processes as a verb. Space can be seen as the process of evaporation to create a new future. Space can be seen as the process of condensing to create a new inside.

The temporal relationships of temporal verbs, as opposed to temporal nouns, are derived from the qualities of light.

The spatial relationships of spatial verbs, as opposed to spatial nouns, are derived from the qualities of gravity.

The past and the future are correlated through the process of releasing cold and evaporating fluid. These processes are complementary due to the fact of their ontological sources. The one process is due to light. The other process is due to gravity. Correlated, they create past and future time.

The inside and the outside are correlated through the process of releasing hot and condensing fluid. These processes are complementary due to the fact of their ontological sources. The one process is due to light. The other process is due to gravity. Correlated, they create inside and outside space.

Releasing cold light and condensing fluid gravity creates the past-inside.

Releasing hot light and evaporating fluid gravity creates the future-outside.

Final Dimensional Recipe

Reserve for Bundal Log Identification

This chapter describes how Bundal sentences compile to describe reality. The concept of a sentence represents condescension to human language. Tasting and sharing a recipe is a more accurate description of how Bundals describe reality.

The final dimensional recipe includes four ingredients. They are light, gravity, time, and space. When seen as nouns, humans describe the ingredients in terms of a substance ontology. When seen as verbs, humans describe the ingredients in terms of an event ontology.

Humans can only SHFTTS the past-inside. In terms of light this limits them to hot. In terms of gravity this limits them to wet. In terms of processes this limits them to exploding, releasing hot light. In terms of processes this limits them to expanding, evaporating gravity fluid.

A place in reality is fully described by a Bundal story. Each story requires only four sentences. Each sentence contains a noun and a verb. Two adjectives modify the noun and two gerunds modify the verb.

DNA has nothing on Bundal stories for its simplicity to create multitudes of unique stories. The Bundal alphabet only has four symbols which can be arranged in a multitude of sequences. Their direct evolution from light and gravity did not require them to mimic sounds limited to their oral cavities. They do not have oral cavities *per se*.

Properly used, four letters are all any alphabet needs. Humans have discovered their two-lettered alphabet, which they call bits, is insufficient for reasons of linearity. Four bits is sufficient for non-linear communication.

The four sentences used by Bundals include the following:

- A sentence with a temporal noun and spatial verb.
- A sentence with a spatial noun and a temporal verb.
- A sentence with a light noun and gravitational verb.
- A sentence with a gravitational noun and light verb.

Each noun is defined by two opposing adjectives.

Each verb is defined by two opposing gerunds.

Each sentence contains six words.

Figure 30: Two species of time and two processes of space.

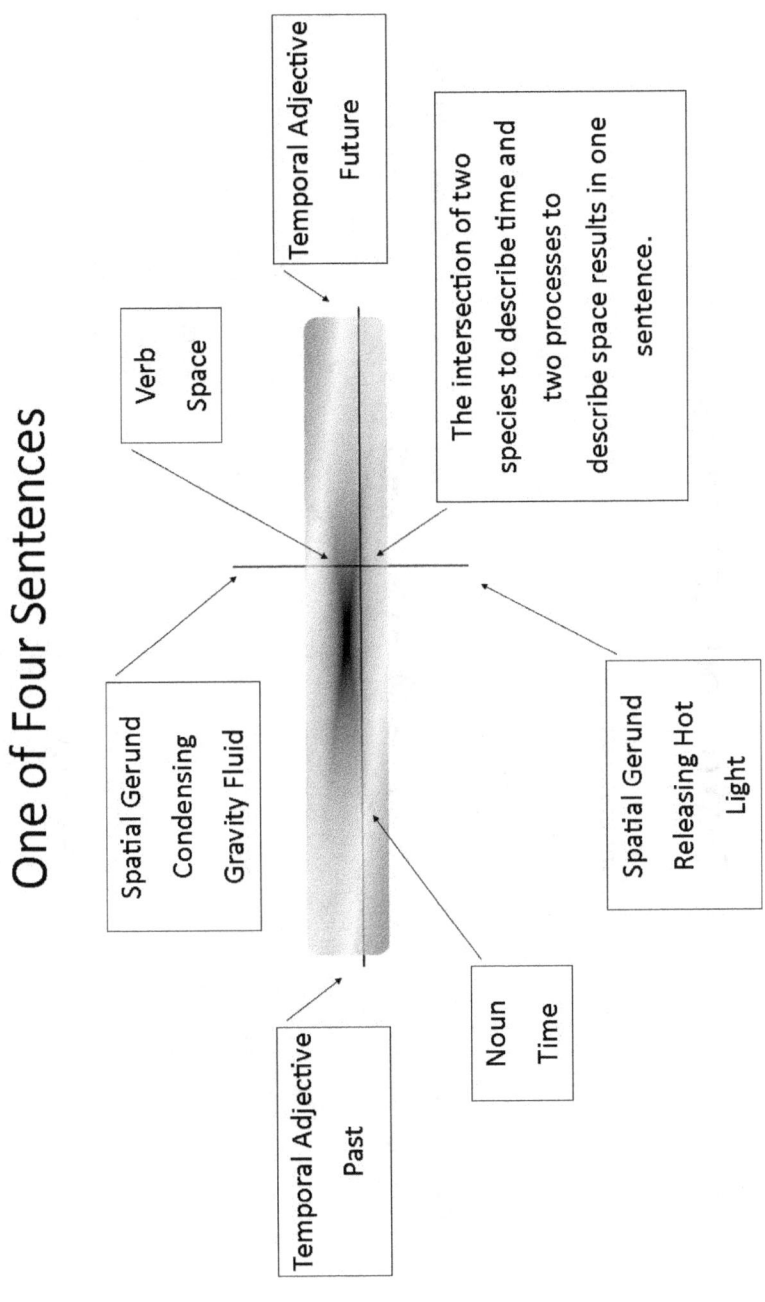

Figure 31: Two species of space and two processes of time.

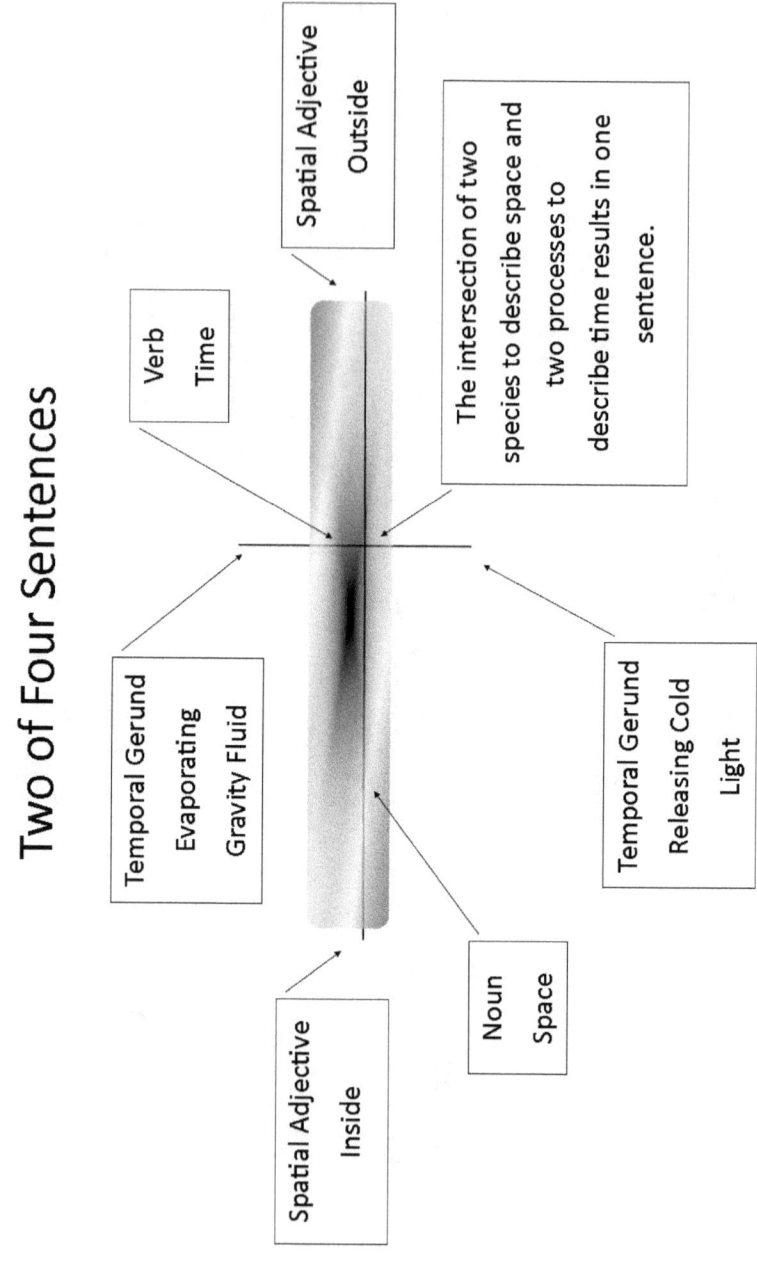

Figure 32: Two species of light and two processes of gravity.

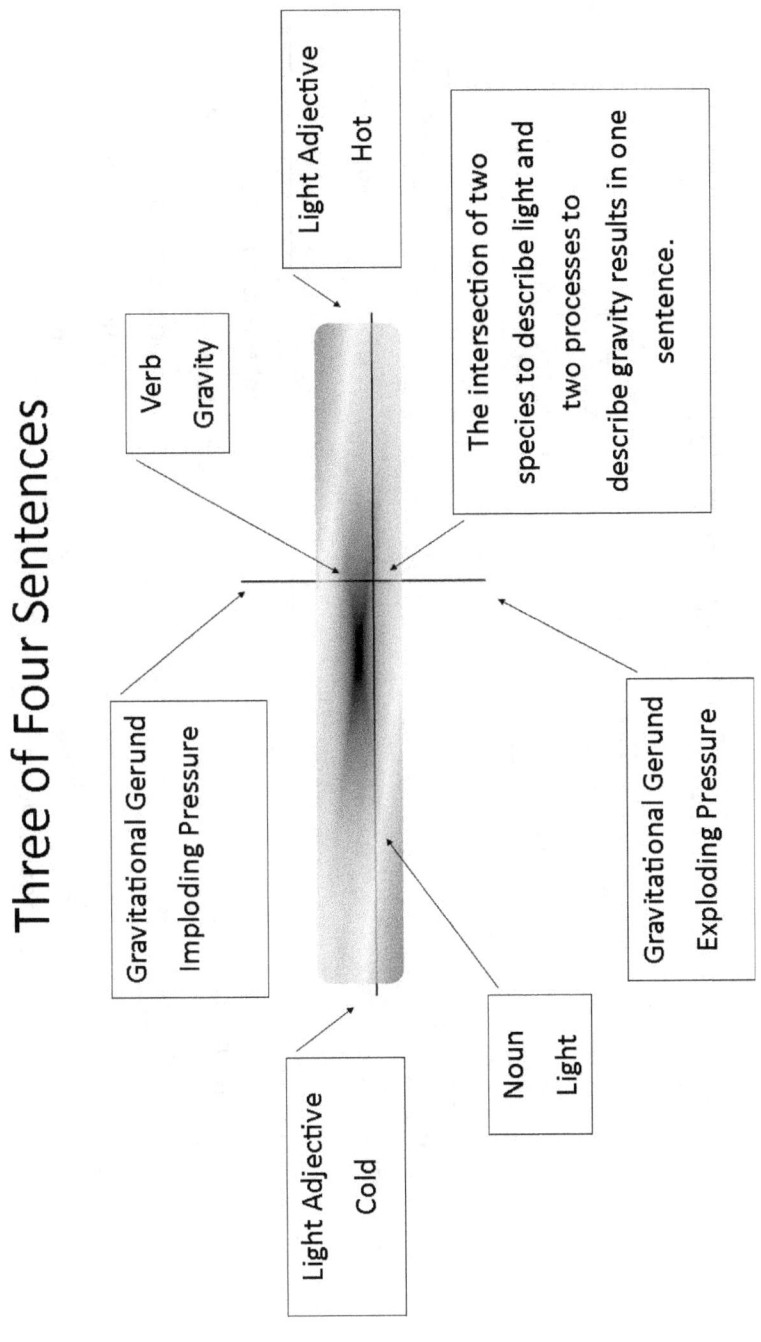

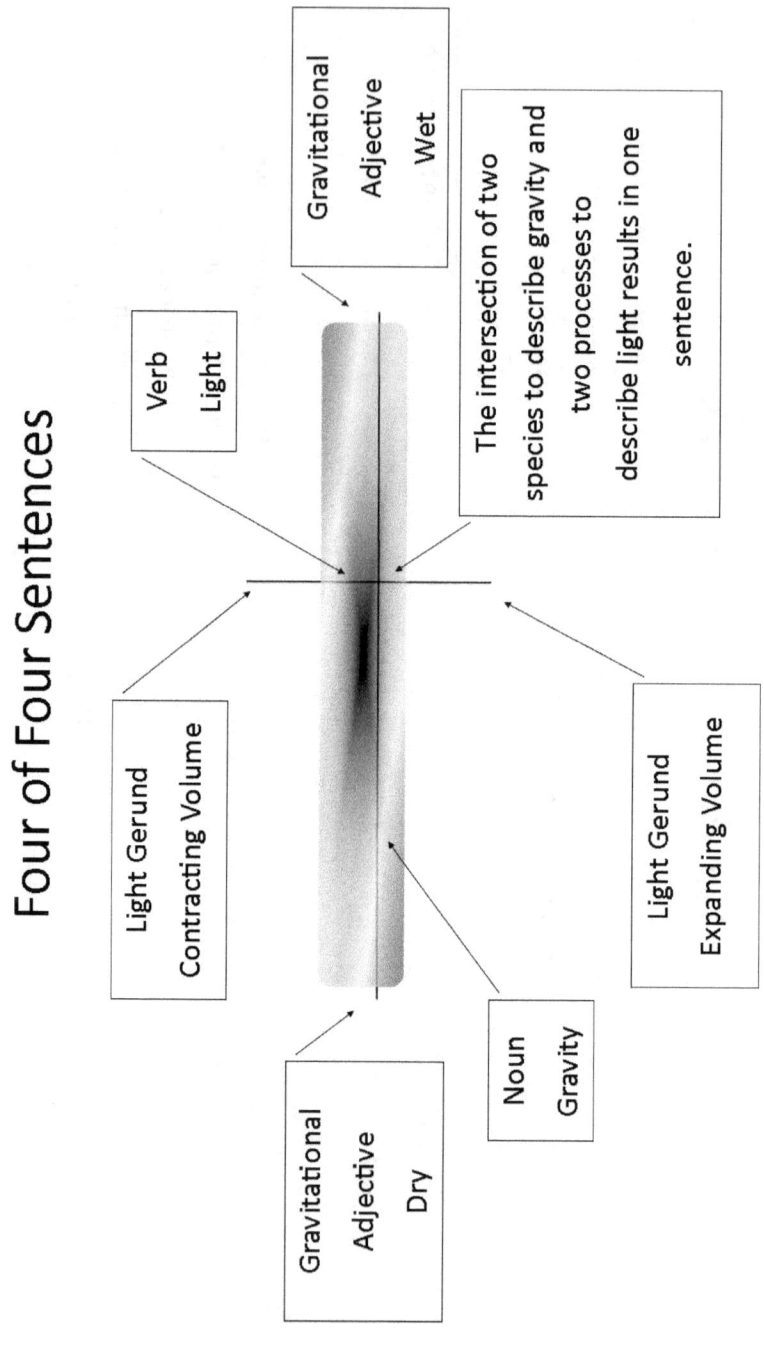

The four sentences come together in what appears to be sixteen dimensions. Time and space have two species each giving rise to the appearance of four dimensions. Light and gravity have two species each giving rise to the appearance of four dimensions. The interchange of nouns and verbs gives rise to the exponential nature of dimensional study. The first ordered dimensions give rise, in our universe at least, to the second ordered dimensions. Four times four, or four squared, equals sixteen apparent dimensions. Humans falsely assume the sixteen dimensions are set in time. All four basic dimensions are set within and without each other.

Figure 34: Fanciful picture of four sentences describing an actor and action within reality.

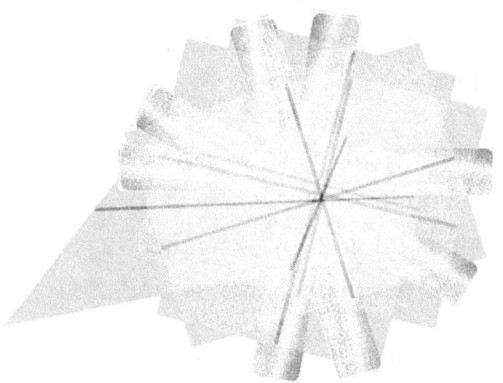

Light and gravity supply the stage. These are the four conditions describing why and where.

- Light as a noun and gravity as a verb.
- Gravity as a noun and light as a verb.
- Each noun is defined by two adjectives.
- Each verb is defined by two gerunds.

These conditions provide the background for the space and time actors and actions, as perceived by humans.

Complementary to the conditions are the notion of length (size), rate (variable time), the wave-particle context, the entropy of matter as a gravitational process, and the entropy of energy as a gravitational process.

The metaphor of a stage implies that relativity is more than just a frame of reference. It has ontological as opposed to phenomenological merit. The frames of reference are the dimensions themselves. Humans might assume that none of the four frames of reference, otherwise known as the four basic dimensions, may be privileged. Bundals know light can serve as a privileged frame of reference within any one information system. Remember that light as information, operating as a frame of reference, has both input and output.

The space and time actors and actions have four conditions describing what and when.

- Time as a noun and spaces as a verb.
- Space as a noun and time as a verb.
- Each noun is defined by two adjectives.
- Each verb is defined by two gerunds.

Humans can only read certain sentences due to their limitation as space and time creatures.

Sentences of measure are past-inside.

Sentences of prediction are future-outside.

Sentence of light and gravity seem quantum mechanical.

Sentences of entropy seem dynamic.

Quantum mechanical sentences and dynamic sentences are very much alike from a Bundal's perspective. Humans tend to SHFTSS them as two different systems.

This implies quantum mechanics and dynamics are both elements of the same sentence. Relativity is rather easily explained once one goes beyond space and time as the only two dimensions.

Observant humans will ask questions about the "who" of Bundal sentences. We cannot answer the question of who by discussing who's. Who can be answered only by what, when, where, and why.

There is no self-proving axiom.

This takes the composers into the realm of religion as categorized by twentieth-century humans. The question of who will not be fully addressed in a recipe for physics. Dr. Suess will have to forgive us. The composers will note the recipes have to come from a who. This subject will be further tasted in *Tubal's Recipe for Religion*.

Notice the possibility for a who in the following visual description of the Bundal recipe.

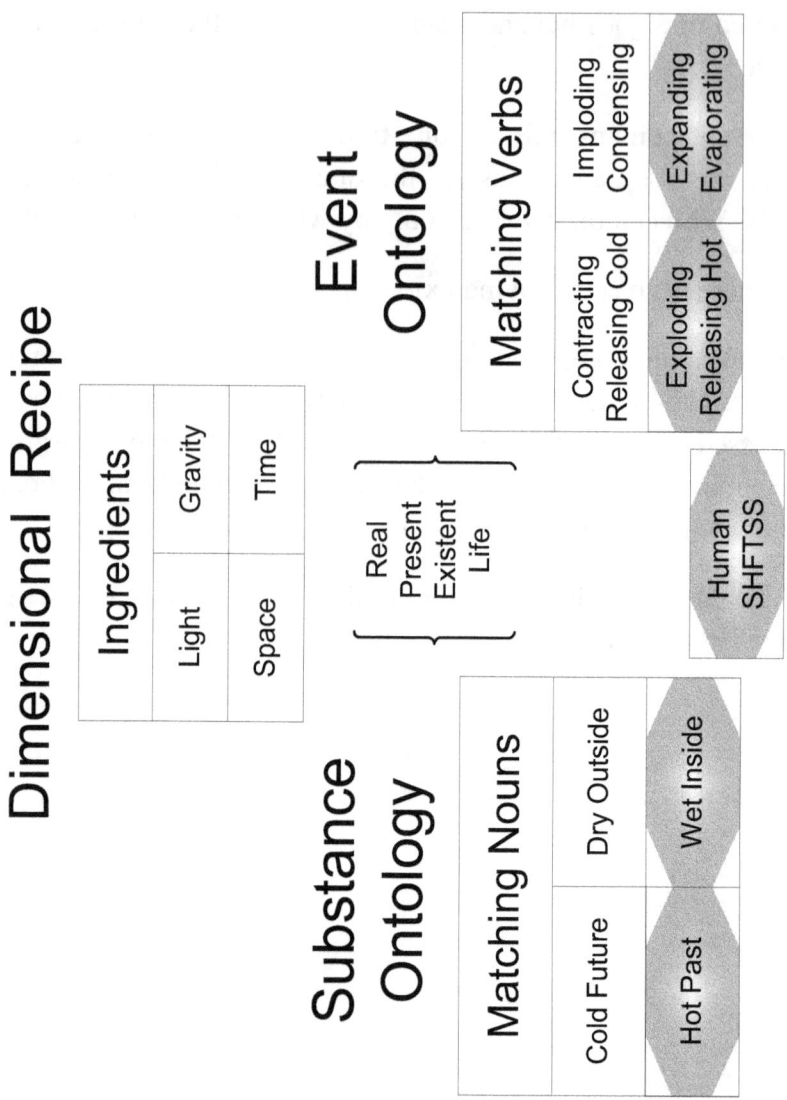

Figure 35: Dimensional recipe for a substance ontology and an event ontology.

There is a space for the real between the inside and the outside.

There is a time for the present between the past and the future.

There is a fluid status for gravity that exists between the wet and the dry.

There is a life status for light that exists between the hot and the cold.

As time and space creatures, indirectly evolved from a drop of gravity fluid due to the calling of light, our ontology is not directly real, present, existent, or life. We are icons of real, present, existent, life.

Human physics deals with ontology, phenomenology, and cosmology. It dismisses the place of God to a dot, or even less.

It sees the hot past, wet inside as the whole of reality.

It sees the exploding, releasing hot and expanding evaporating processes as the whole of reality.

In its twentieth-century form humans physics not only does a disservice to God, it does a disservice to ontology, phenomenology, and cosmology. Its methodology claims to be in awe of the universe it so easily dispatches in favor of the laboratory or mere mathematical models.

The Table of the Gods

The Table of the Gods was considered too offensive to scientists for inclusion in this part of the composition.

The Table of the Gods was considered too offensive to religionists for inclusion in this part of the composition.

The Table of the Gods will be included in *Tubal's Recipe for Ethics*.

Spinning Wheels

The composers have been singing of rotations as an artifact. What follows is an explanation for spinning wheels. Remember, the so-called components of the wheel are temporally separated.

Time and space are orthogonal in either the past or future. Spinning wheels increase temporal dilation in one direction while constricting space in the other. This produces heat in one direction while allowing slippage (lack of friction or traction) in the other.

Worship of Ourselves

There are wonderful human stories about aliens intercepting the Sunday morning broadcasts of American football. The difference of attitudes among the aliens might depend on whether they were from Mars or Venus. Regardless, the stories make comedic the fact that two human tribes, in their own colorful tribal regalia, fight over an inflated pigskin. Some stories indicate that the aliens assume we humans are worshipping the pigskin.

These stories bring a smile to human readers who can see how aliens might misinterpret the importance of the sport. Some aliens might conclude we pay a great deal of money to watch sports.

We are glad to laugh at ourselves.

Many humans are not aware of how some of their more serious stories might sound to Bundals.

> Human civilization was forever ~~changes~~ changed when Galileo, we mean Lipperhey, put a convex lens with a concave lens (shades of Omega) within a tube and could magnify objects three times. Religious and scientific wars were fought over these two pieces of glass.
>
> It was not until 1929 we even knew other galaxies existed. So is 1929 an achievement? Or is it a sad commentary on our inability to see our universe?
>
> It does not take much for humans to light a fire and change the course of the universe.
>
> The smartest humans rolled things rather than carry them.

It is difficult to know whether fire or the wheel rank as the greatest invention of all time. Some claim it was beer. If you look at the history of the human race, the invention of beer seems to coincide with the ability to engage in angry talk. It is followed by a three-thousand five-hundred years ability to write messily.

Is there not an irony that Newton's alchemistry was scientific and his mathematics was artistic?

Human mathematics is an important language tool for extrapolating beyond human phenomenology and advancing our species interest in the rest of the universe. Its importance to us does not imply its importance to other species. What we extrapolate, they may directly sense. At least no known species on earth seems terribly enamored with numbers. Rumor has it that not all humans are terribly enamored with numbers either.

Euclidian Steps to Einstein

If Bundals do not like human math, then what of Euclid?

Human Joke:

Which comes first: The egg or the chicken?

Bundal Joke:

Which comes first: Human math or human reality?

The composers are striving to build a bridge from the human SHFTSS to the Bundal SHFTSS. The sounds of this aria are not those of Bundal steps. They are the sounds of an Euclidian duet heard as a solo by the Einstein-god.

Galileo and Newton brought movement to Descartes. They changed the name of an axis from one letter to another. As all followers of Einstein know, t dominates x or y or z. Even the Latins and Greeks agree on this.

Bundal ethics requires a disclosure. The great insights of Euclid, Descartes, Galileo, Newton, Einstein, and other classical thinkers have been reinterpreted by twentieth-century disciplinarians to fit a twentieth-century worldview. It is not that people long ago were so smart. It is just that they knew exactly what we needed to prove our current thinking.

Which came first: The two-thousand three-hundred year old insight or what twentieth-century disciplinarians say was said?

Just look up what Euclid said about parallel lines and compare what he said to what twentieth-century disciplinarians said he said.

The sacred scriptures of major religions remain the same. They just happen to be reinterpreted to fit our present context. They are so useful because they agree with us.

The sacred insights of the major sciences remain the same. They just happen to be reinterpreted to fit our present context. They are so insightful because they agree with us.

These are the Bundals steps from Euclid to Einstein. (Bundals do not actually walk. They sing and taste.)

Step One
Bundal Log in Pain

The original pitch is established by Euclid. Descartes loved the music. The melody is not disjunctive.

Figure 36: Descartes' space line.

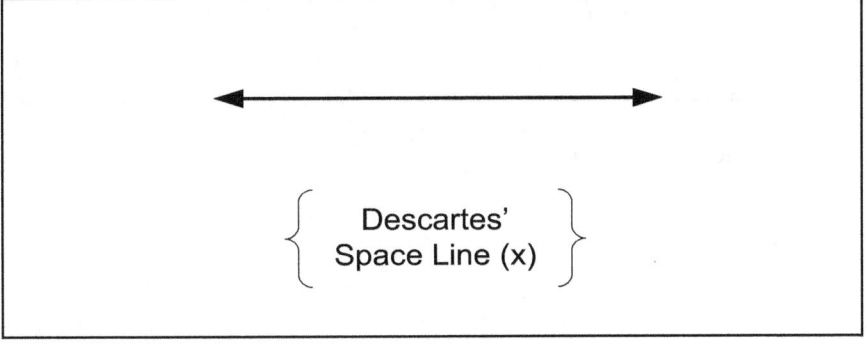

Step Two
Bundal Log in Pain

Newton added time. Or was it Galileo? Regardless, both of them loved Euclid.

Figure 37: Newton's timeline.

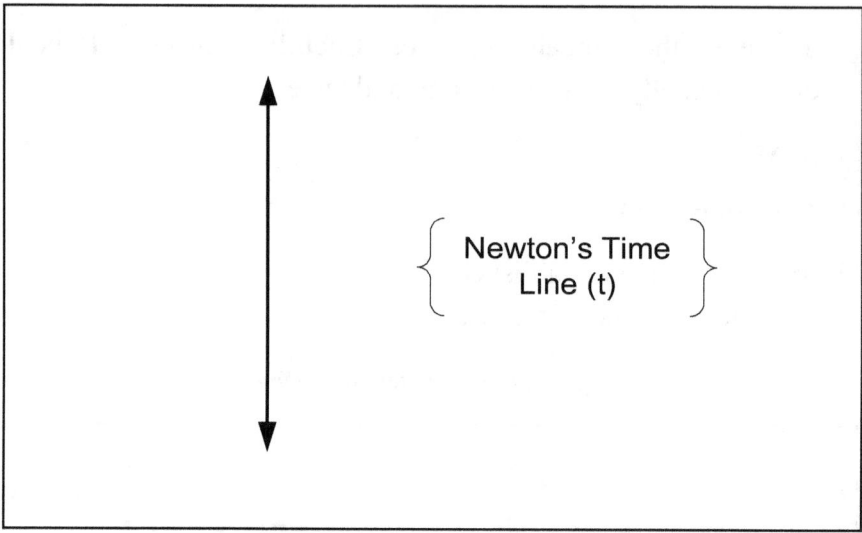

Riemann added time too.

Step Three

Bundal Log in Pain

A careful examination of the following, revised space line implies zero has become nothing.

The Bundals want to know if humans consider this revised space line as one line or two.

Does it represent one set or two?

Figure 38: Revised space line.

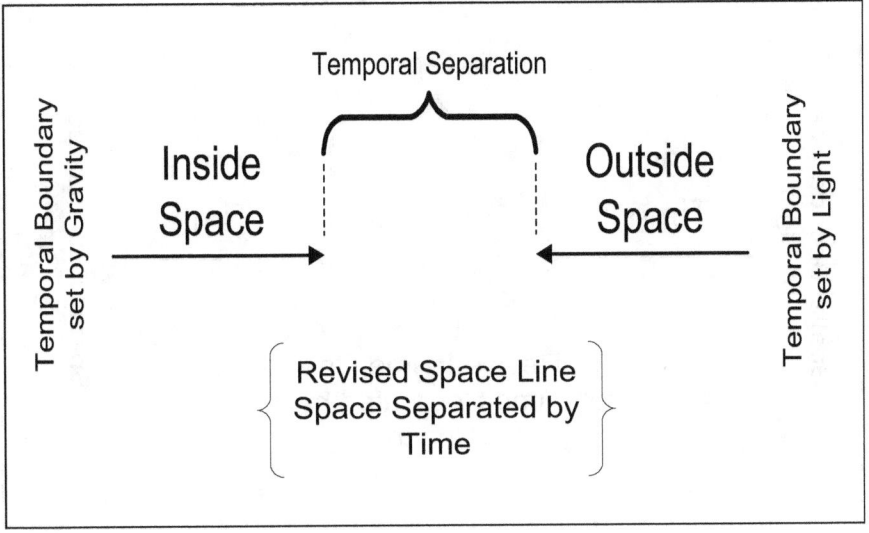

Step Four
Bundal Log in Pain

A careful examination of the following, revised time line implies zero has become nothing.

The Bundals want to know if humans consider this revised time line as one line or two.

Does it represent one set or two?

Figure 39: Revised time line.

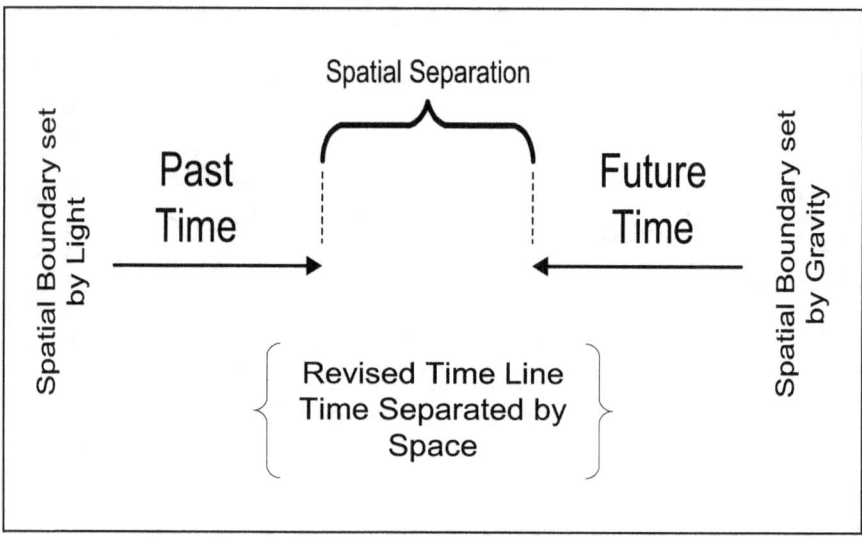

These revisions of Euclid's lines flow together rather than apart, kind of like the magnets of the universe. Notice they do not originate in eternal time or infinite space.

Step Five
Bundal Log in Pain

Lines have always had direction. Just ask north, south, east, and west.

Figure 40: Directional flow of space line temporally corrected.

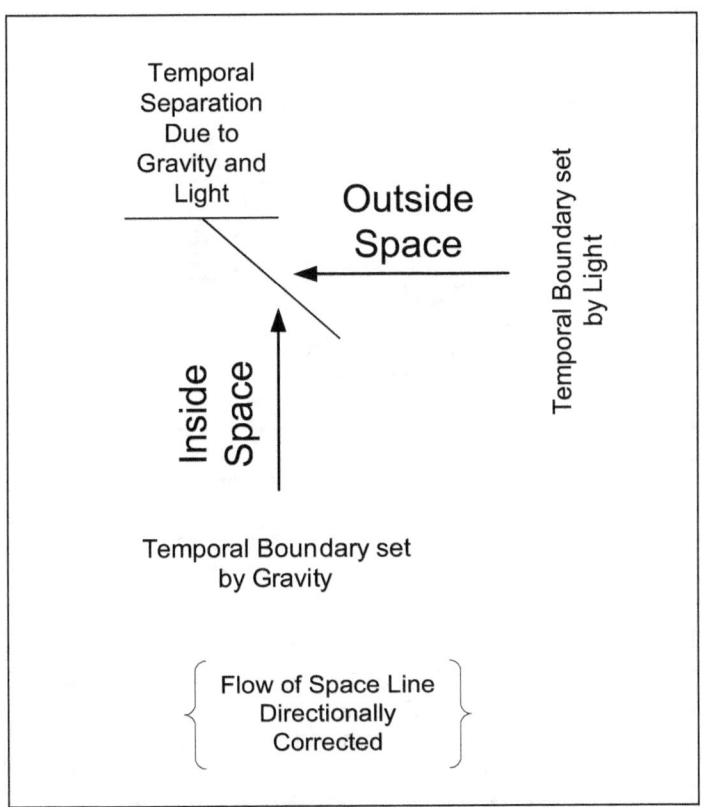

The composers note the universe is simple. There is an initial orthogonal relationship between space and time. Temporal separation is not just spatial. It is directional as well. For the sake of convenience, simplicity, and to remain congruent with our earliest models of flat space, we will assume the limits of space and time are orthogonal to each other.

Lines have always had direction. Just ask north, south, east, and west.

The composers are starting to feel like string players. The steps within the music continue.

Figure 41: Directional flow of time line spatially corrected.

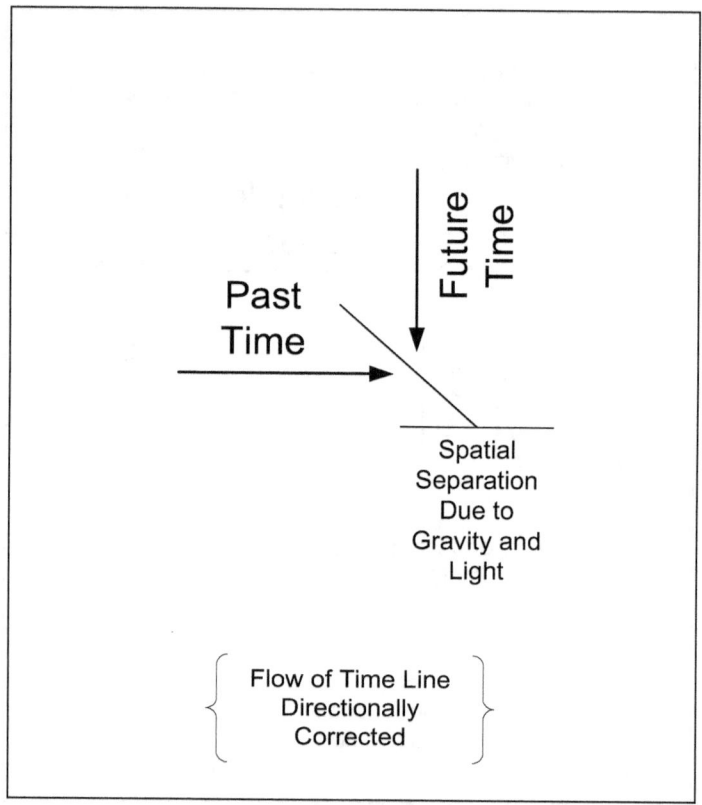

Step Six
Bundal Log in Pain

The Euclidian duet becomes the Einstein-god's solo.

How the Euclidian duets are brought together is a timely matter of gravity and light requiring space to understand. Space and time are insufficient to explain space and time.

Figure 42: Time and space visualized together.

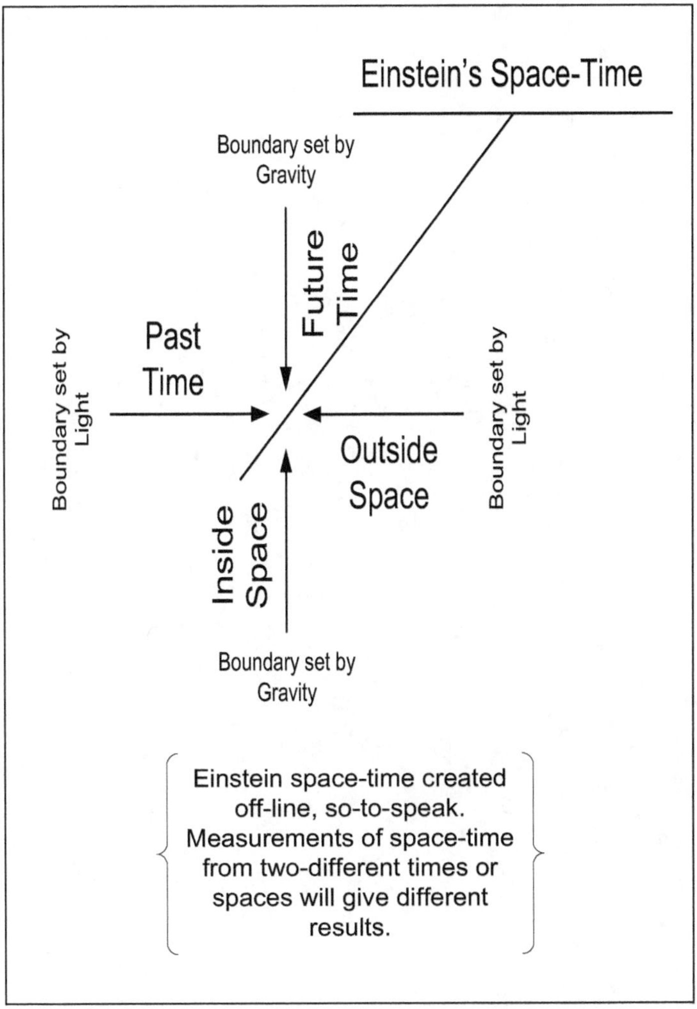

The implications are quite clear. There are boundaries of space and time set by gravity. This differs from the twentieth-century notion that all of space-time is structured or formed by gravity alone.

Less familiar, even strange (forgive the electrodynamic pun), is the notion that the species of time and space are also limited by light. This gives rise to being comfortable with notions of space-time as being on or within the light cone.

Step Seven
Bundal Log in Pain

The solo requires plain music.

- If Descartes' lines can be discontinuous …
- … then does not this beg the question of units?
- If x is one dimension, x and y are two dimensions, and x, y and z are three dimensions …
- … then why can we not have a coordinate axis that accounts for two one-dimensional lines and two two-dimensional planes?
- We broke up the lines.
- We added a second dimension to just two of the four quadrants.

There is a refrain in our composition that clearly states future time and outside space are planar.

Past time and inside space are linear.

Here are four recipes for three dimensions:

- Add x plus y plus z.
- Add x plus y plus t.
- Project x and z through y.
- Project x and t through y.
- Which ones are x, y, z, or t? Label them like you want. We are referring to the projections as the yth dimension.

Figure 43: Two of four quadrants gain a second M-dimension.

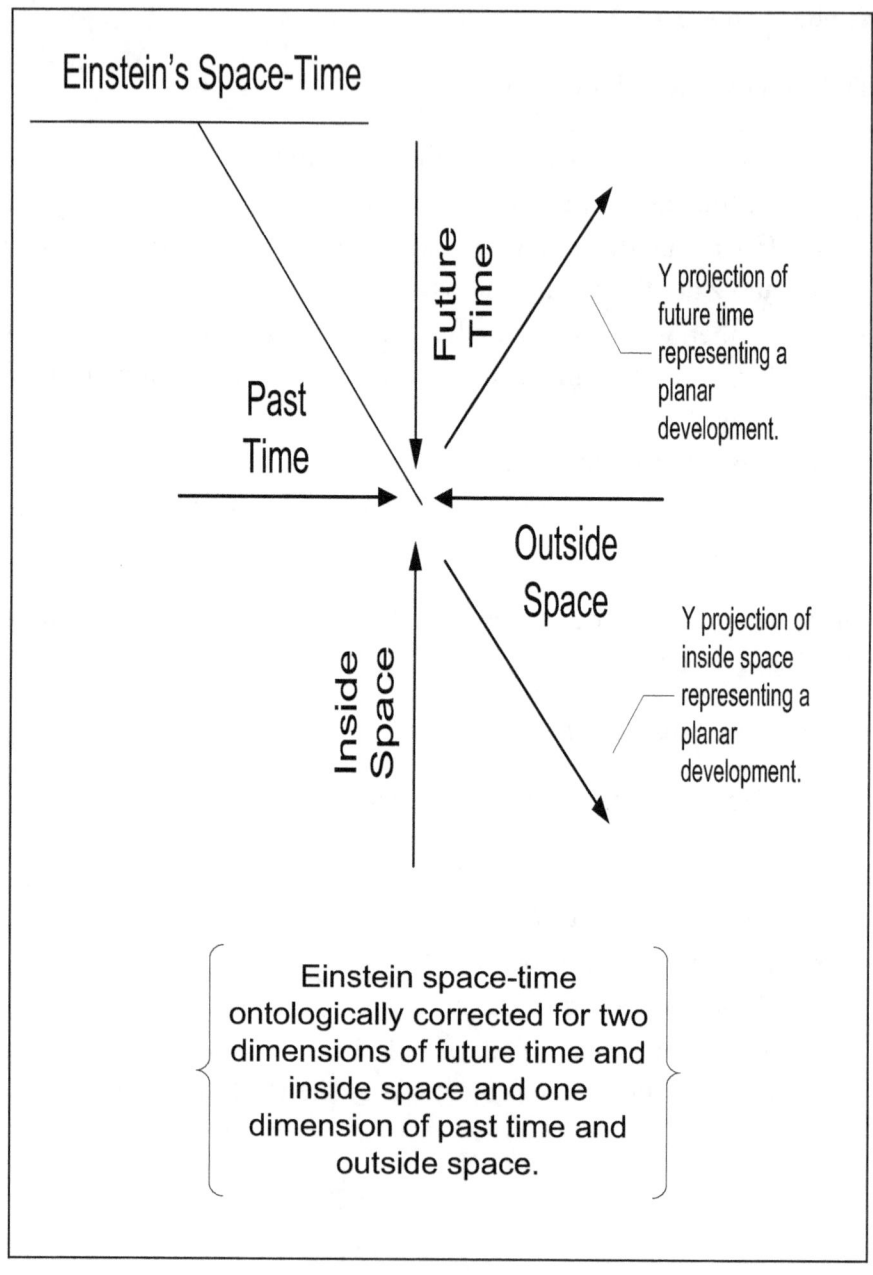

Einstein's Space-Time

Future Time

Past Time

Outside Space

Inside Space

Y projection of future time representing a planar development.

Y projection of inside space representing a planar development.

Einstein space-time ontologically corrected for two dimensions of future time and inside space and one dimension of past time and outside space.

Step Eight
Bundal Log in Pain

You have never seen Euclid or Descartes more tense.

Past time and inside space stay in the same place.

Future time and outside space twist around. They should have had a real fulcrum instead of those breaks in their lines. Maybe they do? The Bundals are letting us sing but they will not tell the composers if they are listening.

> If the y-projection of future time and outside space is the same, then the rotation of future time and outside space could be about their shared y-axis. In this case future time and outside space are contiguous. Their planes share a single line.

Rotation is a mathematical way of describing a relationship between two one-dimensional quadrants and two two-dimensional quadrants.

Figure 44: The Riemann metric tensor.

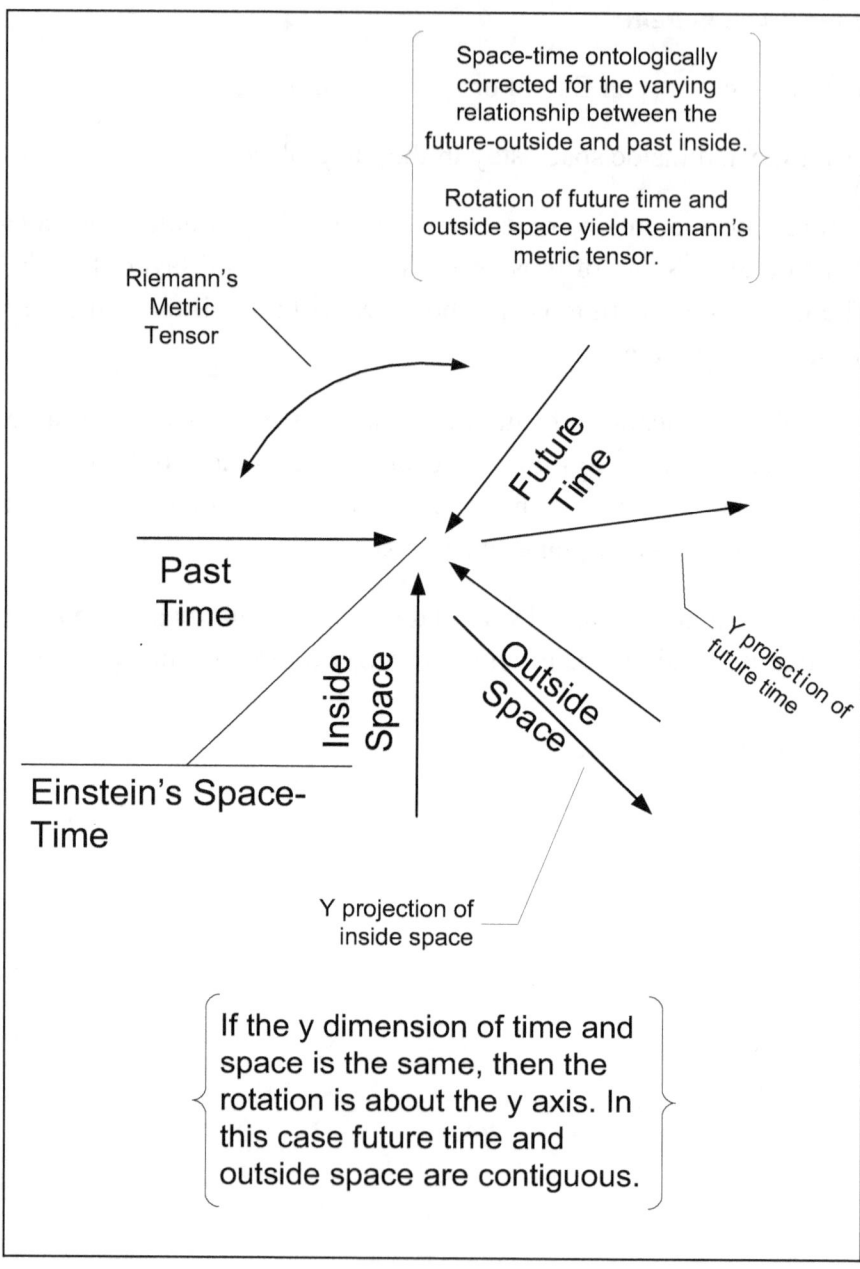

Space-time ontologically corrected for the varying relationship between the future-outside and past inside.

Rotation of future time and outside space yield Reimann's metric tensor.

Riemann's Metric Tensor

Future Time

Past Time

Y projection of future time

Inside Space

Outside Space

Einstein's Space-Time

Y projection of inside space

If the y dimension of time and space is the same, then the rotation is about the y axis. In this case future time and outside space are contiguous.

Hyperspace has to give Euclid a break.

When two humans observe the same event, they are sharing the same inside space and the same past time with the event. They are all locked into place so-to-speak. If they did not share the same space nor the same time, they would be unable to observe the event. Past space is unitary and does not have parts. Past time is fully determinate. This quality of the past-inside enables measurements to take place upon the past-inside. Memory enables humans to compare one measurement from one past-inside to the measurement of a different past-inside. The comparison of the two measurements can go in either direction depending solely on the priority given to older ~~memoires~~ memories over more recent memories. The priorities are set by human standards. This comparison of two measurements explains the apparent bi-directional flow of time.

> The measurement is just as valid if the broken egg jumps off of the floor and reassembles itself on the table top.

Our two humans are observing the event from different perspectives due to being in a different outside space from the event. They are also observing from a different temporal perspective due to the potential nature of future time. In a sense, their perspectives are rotated from each other from the place and time of the event.

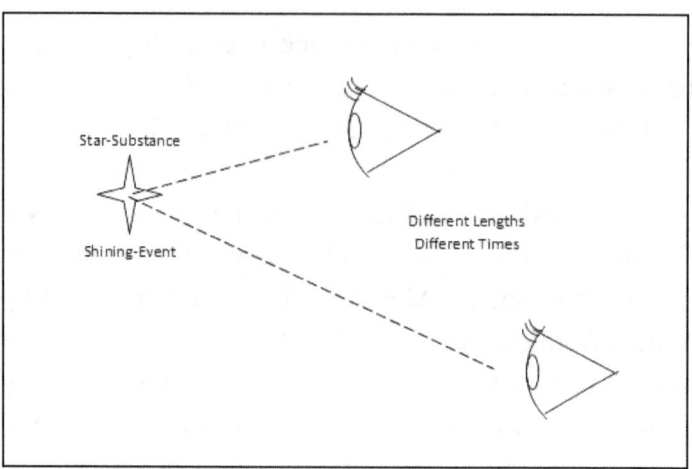

This simple eye-chart is classically used in optics, math, geometry, and physics. It is functional only from the outside-future. Humans in different places and times do not SHFTSS each other. They must share the same space and a time in order to SHFTSS each other. The space and time human observers share with an event is the past-inside.

It is difficult to discern temporal projection through two-dimensional space with only one eye. The reason for two eyes is to perceive temporal differentiation. One eye can see a space as easily as two. The timing difference requires temporal separation of the observing sense organs.

No measurement can be made until the potentials become actualized and are made part of memory.

It is not that space and time shortens or lengthens or slows or speeds *per se*. Space and time only SHFTSS like it shortens or lengthens or slows or speed due to the observer's difference in outside location of future time, a difference that draws stark contrasts when measuring a fully determined place within the past-inside.

One way of conceiving this is the farther my outside is away from the inside, the smaller my ruler must become I order to measure you.

One way of conceiving this is the closer we get as our future potentials move towards colliding, the more our hearts begin to beat as one. Our time zones are heading toward a unity rather than a difference in timing.

The eight steps of Bundal music function as a condescension to the human notion of an octave.

Brahmasputha Siddhanta

Reserve for Bundal Log Identification

Notice the Bundal number line. (Arrrrrghhh- Sorry. The Bundals insisted the composers insert their pain here.) They constructed a number line for the composers.

Figure 46: Bundal number line.

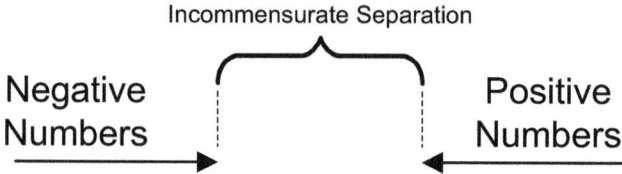

Their number line honors Pingala's *śūnya* or void.

Why do you need zero or infinity to measure? You measure nothing with zero.

There are no zeroes on the Bundal number line.

There are no infinities on the Bundal number line.

Incommensurables give rise to irrational numbers. Who wants to be irrational?

Irrational numbers are used for art. Just ask the Egyptians.

Incommensurables are caused by inappropriately mixing spatial and temporal units. Rather than ascribe mystery to reality, why not question the measuring system?

Schrödinger's wave function of an electron encodes information about a three dimensional wave. Unfortunately, every time an electron is added, Schrödinger adds three more dimensions to the wave. This makes the periodic table an effective chart for listing the number of dimensions necessary for each element.

Who needs string theory to multiply the number of necessary dimensions when you have a periodic table?

Chemists have missed their historic opportunity for leading physicists in the number of dimensions necessary for the existence of life. Carbon requires eighteen dimensions.

This is a Bundal ruler made for humans. (Bundals do not use rulers.)

Figure 47: Bundal ruler.

This is a strand of human hair to measure. Human hair comes closest to what Bundals believe humans know about strings. Mathematicians need not be insulted.

Figure 48: Strand of human hair.

The Bundals are confused by humans who insist on straightening the hair out to find its length. Does it not have the same length regardless?

Why not change the length of the hair to be commensurable with your unit measure?

Who can measure a littoral zone without time?

This hair is four whites (arrrrghhh) and three blacks plus some black long.

Notice no zeroes were needed to measure the hair.

Notice no infinities were needed to measure the hair.

Numbers were used as an attempt to understand human language.

This is where humans would scoff at the Bundal measurement process. But remember, we use to measure the height of horses by the length of the King's hand.

The Greatest Confirmation of the 20ᵗʰ Century

The frog without legs went deaf. He could no longer jump on vocal command.

His name was Sir Arthur Eddington. This disciple traveled afar to photograph a solar eclipse on May 29, 1919. If the Einstein-god was correct, then starlight would be bent by the gravitation of the sun. This would make the star appear on the photographic plates in a different location than its actual place.

Eddington confirmed the universe of the Einstein-god. We are told this confirmation was based on solid scientific principles in spite of the following.

- The poor quality of Eddington's photographs.
- The contradictory findings made simultaneously at Sobral, Brazil which supported the anti-Mercurial Newton over the Einstein god.
- The Einstein god had changed the mathematical outcome for his prediction by 1919. Ask a string theorist. The more predictions the better. One of them is bound to be correct.

Tainted evidence triumphed over belief. Eddington confirmed the greatest discovery of the 20ᵗʰ century.

The Bundals were ecstatic. This was objective science as good as humans can make it. Maybe now the humans would discern the power of light to create and control gravity wells.

Composers sung and will sing:

Eddington confirmed the existence of a third dimension beyond mere space and time. The astonishing result is that simply bending light distorts space and time sufficiently to create great depressions that trap baryonic matter. Our sun expends great energy attempting to break out of its light-prison. The plasma creature floods space and time with countless photons in a vain attempt to eclipse the bending light. Ironically, its photons would bend in their attempt to escape, trapping planets and moons. If only it could expand beyond its imprisonment and fill the solar system with its baryonic matter.

The light dimension, confirmed by Eddington, refers to hot light and cold light. It controls gravity.

This dimension was once profanely referred to as ether. It was demonstrated by Newton's vortices. And Eddington proved this phase-state dimension held a most powerful form of matter and energy within it. This white and black form of matter and energy created a prism effect in space and time that bends light to control gravity.

Minkowski and Kaluza-Klein had admitted extra dimension were possibilities. Human scientists became even more fixated on space and time as privileged dimensions, even combining them as Hyper-space, instead of distinguishing between the spatial and temporal.

Michelson-Morley failed to understand that light, which flows along the path proscribed by the additional dimension, would flow at the same speed regardless of rotation. Simply spinning baryonic matter around, within baryonic matter, and measuring it by a space and time formulation, would yield no evidence of an additional dimension. There was no space-time ether for humans.

How long would it take the humans to understand?

- Light is not the best way to measure most things. Light is so powerful it will intrude upon most processes, literally causing the process to collapse at the point of measure.

- Space and time are secondary dimensions to the primary dimensions of gravity and light. Pursuit of space and time measurements will shrink the perceptible universe to less than five-percent of accountable mass and structure.
- Smell is the best way towards discerning the gravitational dimension. It is an accurate measure of the processes of white and black energy and matter.
- You cannot measure like by like or light by light. You will always be asking the question, whose light should we use? How many hands high is a horse? Which King's hand should you use?
- Measuring like by like creates paradoxes in number theory. You end up with zero, or infinity, or irrational and complex numbers. Worse, you come to believe one set of infinities is larger than another set of infinities.

God said, "Let there be light." Light rules the universe, controls gravity, and it rules space and time.

Bohr's Religion

How much power does it take to be powerful?

Niels Bohr did not need religion.

He had one.

It was called the electron.

It jumped.

It teleported

It does not exist until you measure it

I wonder what Freud would have said to such a patient?

Why do you need a real God if you are going to establish the existence of a real quantum world?

EM is brought into existence as its measuring ruler.

At least Bohr was better off than that other fellow, Einstein. He never could figure out God's game. Einstein believed it did not involve dice.

Dice are used in many games. Many of these games are designed to be won by the one with the most skill. I guess the dice are used to let you know whether you are going to win quickly or slowly.

You do not need to load the dice to win.

Einstein's greatest blunder was not the cosmological constant.

Einstein's greatest blunder was not the repudiation of the cosmological constant.

Einstein's greatest blunder was he did not understand the rules of the game. God does play with dice and will always win.

You do not need to load the dice to win.

You just need some light information.

If only we knew God's game.

For example, look at money.

The world's economy uses information for its currency. Money is the mere by-product of the value of the information you have and how you use it.

Who you know, what you know, when you know, where you know, and why you know equals information that can be used to make you or your friends wealthy.

How little information does it take to become wealthy?

Not much, if the timing is right.

Does knowing the right person sufficient for wealth generation?

Sometimes.

Does having inside information, no matter how derived, sufficient to make one wealthy?

Sometimes.

Does knowing how something works when few others in the world know the same thing sufficient to make someone wealthy?

Sometimes.

How true does the information have to be to make you wealthy?

I do not know.

Fiat money and math have a lot in common. If people accept your mathematical equations as being truly predictive of nearly anything, you can make a lot of fiat money.

How little information does God have to know in order to win?

The correct answer is, "just enough."

You do not need to be "All Powerful" to get your way. You only have to be powerful enough.

You do not have to be able to kill Achilles. You only have to cut his heel in order to win.

Supposed I had a one-billionth of one-percent certainty I could create an atomic explosion by simply changing the trajectory of three atoms in the atmosphere. If I could influence the trajectories of trillions of atoms every Planck-second, how many atoms would I have to maneuver to get an atomic explosion? The final number depends upon my patience. And God is a very patient being.

Now you know why it took four hundred years to get the people out of slavery. This may seem an inordinate amount of time for human slaves, but whatever God says will happen … will …

… eventually

… sooner or later

…it is a done deal.

You do not have to have absolute control and/or assurance of every outcome in order to still be God. All you need is patience, persistence, desire, and light information.

What about prophecies?

> Every one of us has a chance to be used of God to fulfill some of them.

God does not even have to know what particular players are going to do in order to win the game.

> Just guess well and respond appropriately as the game progresses.

> Or have a lot of patience to wait the other player out.

>> Bundals consider us clueless concerning our lack of innate recognition of reality's informational structure otherwise known in the Bundal model as light. Without information, how can you exist?

How much information does it take to rule the universe? Just a little properly applied. The converse is also true. How much information does it take to destroy the universe, let alone a planet? Not much.

A little change in the gravitational constant of the universe could go a long ways to earning many worshipping entities, even those that do not like you. All it takes is the right piece of information.

The question is not whether God plays with dice or not. The question is, "What are the rules of the game?"

> Western philosophers are quite causal as they pontificate on the qualities of living in a past temporal time zone.

> Eastern philosophers are quite laissez-faire as they seek to personify the qualities of just letting the future come to us.

> Both groups of philosophers are correct but only when both strategies are seen as evidence of humans living in two, separate temporal dimensions.

The essential question to ask ourselves is, "Do we really lose if God wins?"

This is as scientific of a question as it is religious, theological, and philosophical.

Bundal Joke

Since the essence of understanding other cultures is being able to heartily laugh at their jokes, the Bundals tried this joke out on the composers.

> A physicist, a chemist, and a witch-doctor walk into a laboratory.
>
> The physicist analyzes reams of data only to conclude the data is inconsistent with expected experimental outcomes. She concludes the data has been corrupted.
>
> The chemist drops the vial and ruins the experiment.
>
> The witch doctor dances around his patient and the person feels better.

That is the end of the Bundal joke. Get it?

The composers surmise the Bundals are asking, "Which of the results was scientifically reproducible?"

> It seems the witch doctor has many healthy patients leaving the laboratory.
>
> The physicist and chemist have corrupted or ruined experiments.
>
> Yet humans see the poor witch doctor as scientifically inauthentic. The physicist and chemist claim scientific success since they have at least learned something from their mistakes.
>
> Humans see the poor witch doctor is seen as emotionally manipulative, as if emotions had nothing to do with reality.

In spite of all the objective evidence of most people getting better under his care, and able to leave the laboratory, he is considered the fool.

Worse, the people who believe in him are considered fools too.

The physicist and chemist are held as heroes even though they will be proven wrong in future years.

Bundals know twentieth-century humans are not smarter than their ancient parents. Our ancient parents believed in the religious hooliganism of the tribal priest. Our more recent parents believed in certain philosophical ideals as the penultimate achievement of our universe. Twentieth-century humans believe in the practitioners of physics and chemistry who tell us what to believe.

Next thing you know humans will be predicting the weather.

Has scientific hooliganism replaced religious hooliganism?

Humans might ask, "How many physicists does it take to burn out a light bulb?"

Babel's Bridges?

The composers offer you an inside look (forgive the pun) on op-narrative.

Describe the color red to a blind person. Make sure they know what you know.

After you have spent years teaching the blind person about the different colors, can they pass the test?

How many words for snow does it take to understand snow? How many words for snow are there in the Uralic languages? Can you understand snow if you lived on Venus?

How can you look for life on other planets when you cannot even properly define its existence on earth.

What is a word?

Is the English word, clubhouse, one or two words?

What is a Rankul? Is it a made up word or is it a bamboo tube used as a tool in Indonesia?

How do you communicate between species?

Language?

Whose?

Touch?

Odor?

Does swimming with dolphins make you one of the pod?

There are many lingual techniques used to turn human skulls of gray mush into sophisticated, organized, information retrieval systems. Repetition is great for indoctrination and effective propaganda. That is why some humans do not know how to do math. They were never indoctrinated.

If humans have so many problems with language, how do the composers build a bridge to human beings?

There are in the Bundal log.

There are gaps in the memories of the composers.

Humans cannot even write first thoughts. As soon as they attempt to write them, they have become second thoughts.

How do you reach the human race? Writers are supposed to have a target audience in mind to whom they direct their written thoughts. Which humans should be targeted?

God seems to favor using weak and ineffective people to spread messages effectively.

How does an author morally pick a subgroup as an audience? Can you imagine living in 1935 and innocently picking the Nazis for your first paper on atomic energy?

Bundals know it is immoral to focus on a target audience. There are many things any one audience does not want to SHFTSS.

I cannot make you understand me.

You will have to do some work in order to come over the dividing waters to reach me.

You are the bridge.

Do not blame the Bundals for the ideas in your head. Do not blame the composers for the ideas in your head. Do not blame the author for the ideas in your head. Do not blame your DNA for the ideas in your head.

Man up. Woman up. Indeterminate up.

Twentieth-century science has advanced to the point of not knowing how to accurately describe the difference between a male human and a female human. Just ask any sports association if some of their competitors are male or female.

The notion of gender differences is seen as a ferocious inefficiency of space and time according to Bundals. Bundals SHFTSS three facts (arrrrrghhh) about gender differences.

- Females do children while males cannot do children.
- Males do females because females cannot do children.
- Whether males and females can do children is indeterminate.

All three facts are true.

Babel's Refrain
Reserve for Bundal Log Identification

The bridge you build will lead you towards a new science-fiction.

- Your new world has two discrete temporal dimensions. They are identified narratively as past and future.
- Your new world has two discrete spatial dimensions. They are identified as inside and outside.
- Your new world has two discrete light dimensions. Light is identified as black and white. This is not off and on. This is

more akin to the concept of ingression and egression. Both black and white are proactive.

- Your new world SHFTSS gravity as a fluid. Some humans may prefer to use the concept of phase transitions. Dry fluid and wet fluid are both proactive. They give rise to space and time.

- Your new world has four dimensions inter-related ontologically.

- Your new world informs humans of their phenomenological limitations.

The only thing missing in your new world is a present, substantial, colorful, thinking reality. (See Figure 35.)

This lack is no big deal in a book on physics. Human physics works best without present, substantial, colorful, thinking realities.

Religion is a sham without God.

So is science.

Whether the bridge is worth building is up to you. You will have to do your part.

Your reward will be pieces of the whole that satisfy and challenge you to your core.

What looks like op-narrative to you is a high-entropic informational structure that will lead to areas of low entropy and high value to a Bundal.

You will be able to imagine two temporal dimensions.

Babel Religion
Reserve for Bundal Log Identification

We laugh at a people attempting to build a tower to heaven. God did not laugh.

> The Lord said, "If as one people speaking the same language they have begun to do this, then nothing they plan to do will be impossible for them," (Genesis 11:6 NIV)

You do not have to have a true religion in order to accomplish great things.

You do not have to have a true science in order to accomplish great things.

Credits

No Reserve for Bundal Log Identification

A Description of the Aliens

Intertwining strings are difficult to unknot.

Picture of Bundal

The composers are not allowed to tell you too much about our alien friends. But in this competition of species, what do you expect? We will tell all we know to our fellow human beings.

The Bundals look like a bundle of tubes. Imagine an octopus without a head. All of the legs swirl around into each other. A few have what appear to be suction cups. Other legs appear to have hair. Many of the tubes look like melted PCV pipe.

The tubes form bundles that are somewhat elongated, about two and one-half meters in length, with a diameter of approximately nine-tenths of a meter. The pulsating tubes can and do rearrange themselves for locomotion. There are no hands or feet. Some of the tubes can separate and grasp and pull and encircle and do almost anything a human hand can do. Some of the more muscular tubes can unbundled, and become a series of fine delicate tubes that can manipulate small objects.

The Bundals claim they evolved from strings. Strings are a product of gravity fluid.

Amazingly, the Bundals claim they are "it." There are no other life forms on their original planet. Each Bundal functions as a self-contained ecosystem unto themselves. Each set of tubes harvests and eats the minerals and chemicals that form their world. The upside of all of this is humans will not be in any recipes for Bundal cookbooks.

Bundals would be interested in some of the delicious recipes we use to create what we call deep-space satellites. Our satellites, especially the plaques, are delicious.

Bundals believe the sharing of resources is what brings species together. Friendships are developed over shared meals. They did like our satellites.

There is clear evidence of earlier versions of tubes existing on their planet. But there is no other living creature on their planet, not even surviving little strings that gave rise to the first tubes. It seems each period of their evolution led to the extinction of whatever existed beforehand. They simply self-evolved over time, ever improving and adapting to the changing environment of their planet.

Their evolution is not nearly as complicated as ours. Theirs is disjunctive. Ours seems to be continuous. The Bundals SHFTSS us as merely changing from inefficient species to inefficient species.

You can imagine how surprised the Bundals were to discover there were other forms of life in their universe. This did not really bother them too much since they knew they were the superior form of life. Although they do not understand their primordial origins, they do know they are closest in efficiency to the substrate of the universe sustained by strings. All other forms of life seem to have significant amounts of intermediate steps and processes that are required for them to live. These inefficient forms of life need other forms of life to exist. Bundals are quite convinced that human evolution has taken humans farther and farther away from the purities of relationships found in constituent matter.

The Bundals were really shocked to discover we had evolved as creatures of space and time. The Bundals consider this an indirect form of evolution. They do not even SHFTSS it as evolution, since space and time invariably lead to high entropy.

Bundals are directly evolved from light and gravity, the first-ordered dimensions.

The Bundals do not have organs so much as they have systems. Each system is composed of tubes that wind throughout each Bundal. Some tubes are clear, others have color. You can see fluids flowing through most of the tubes and some of the tubes squeeze and release in rhythmic fashion. Some of the fluids clearly contain particulate matter.

The Bundals shoot liquid and crystals from some tubes to defend themselves or resolved arguments of priority. The liquid does a really nasty job on other Bundals while the shooting crystals are used to puncture an adversary's tubes.

Of the various tubes that form the Bundals, one system has priority over all other systems. It is the system that delivers repair materials to various parts of the body. Bundals actually grow temporary pieces that work like organic robots to repair and replace damaged tubes and systems. Their immune system is the most important system of all.

The Bundals can die but the constituent parts can be transplanted to other Bundals, especially the younger ones, to help them grow or rejuvenate.

Bundals breed by intertwining. Not even the Bundals themselves known the minimum number of partners needed before a new Bundal is born from within the intertwined Bundals. It seems that many pieces of tubes, from multiple Bundals, have to come together in just the right position to create a new and healthy bundle. Among active intertwiners, a new bundle starts to grow every thirty or so earth years.

Some of the tubal systems seem like a glorified refinery. In fact the tubes have a burnt smoke smell, like they are exhausting gas and other by-products.

The Bundals do not have a brain. They operate with an information system, which works in conjunction with the chemical system described above. It is a series of complicated, microscopic sized tubes that appear to deliver information in the form of billions of microscopic pieces that float in a fluid. These pieces can fit together in many different ways and combinations. Over time they eventually connect to create structures of materials that seem to block or allow chemicals to be released into other tubes. These structures appear to form by trial and error and grow a network of bigger tubes, so-to-speak. Successful structures enable the right chemicals to combine without destroying the structure. It seems the tubes can remember which structures function best, and can squeeze themselves to give rise to the greatest number of puzzle pieces coming together to form ever more complex functional structures.

Problem solving grows with each success. It is not so much the transfer of electricity as it is the transfer of light. It seems strange to humans to distinguish the two.

The Bundals have yet to deduce whether the structures of these information systems are self-aware, since they work in such symbiotic fashion to produce and carry information. Does the information have a hidden source? Bundals think not since God is one source of information for light directing gravity.

The Bundals are most sensitive to pressure changes. The wrong kind of pressure will collapse and hinder many of their tubes. The Bundals find living on earth difficult, because of rapid pressure changes in air and water.

The Bundals SHFTSS all sorts of environmental data. Some tubes sense light, others sound, others temperature, and so on. All senses have both inputs and outputs.

Their sense of taste, so important to them, is a partial combination of our sense of taste, smell, and feeling through our skin. Put these together and you have a sense of Bundal taste.

Bundals communicate through forming patterns of color on a communication tube. Whoever can receive the colors on their electromagnetic reception tube understands what is being said. Because of multiple tubes, standard communication is almost simultaneous. Their quick system creates problems of interpretation. Corrections require multiple conversations with multiple Bundals in order to sort out original intent. Words are seldom used.

God
Reserve for Bundal Log Identification

The composers hesitate to share with you whether the Bundals believe in God. What difference would this make to humans whether God actually exists or not? For what it is worth, the Bundals know there is a God. Comedy, on their planet is not about sex. Intertwining is far too difficult and time consuming to make jokes. The core of their comedy centers on stories and situations that assumes there is no God.

Bundals laugh at almost anything that hints there is no God. Actually, several of their tubes pulsate and they let off extra smoke, when they laugh. They think humans are the most hilarious species in the world, and feel comfortable in approaching us due to our sense of humor. They would never imagine a species without God.

Spaceships
Reserve for Bundal Log Identification

Bundals find it remarkable that humans are trying to build little ships to go into outer space. Instead, they did the obvious. They simply stayed on their pressure regulated planet, with all of its resources, and allowed their sun to drag them, and a few other protecting planets, through the universe.

What is easier?

Wasting precious planetary resources or controlling your sun?

History shows they cared for their one and only spaceship very well.

What is a planet but a wandering one?

Since Bundals assume other advanced species would float their planets through space, they have searched for other wandering stars in earnest. They have yet to find any that has the dry, dusty environment and pressure requirements so necessary for string formation.

Over population is not a concern. Youthful Bundals do most of the intertwining. Older Bundals are absorbed in their work and place in the Bundal's mission of simply SHFTSS-ing the universe. In the past, when they needed more Bundals, some of the Bundals mandated forced intertwining. This raised ethical concerns about creating new Bundals. History showed it was best to leave intertwining as a voluntary activity.

How did Bundals find us?

It turns out they broadcast what we would understand as quantum fluctuations that are variously entangled with one or more Bundal information systems.

As part of the universe it is more than subspace. It is the creation and manipulation of time and space itself. Humans are looking for activity within time and space to discern the existence of other species. Bundals look within light and gravity to find other species. Space and time are mere expressions of gravity and light.

One such fluctuation apparently hit a composer's tongue, leaving a bad taste in their mouth. The composer then shared this bad taste with a fellow composer (do not ask for specifics) and the Bundals were then able to connect with both composer's conditions.

Both composers now sense a series of sub-atomic strings lining up and coalescing to form what they SHFTSS. The composers SHFTSS a connection to something that they are not even sure is there.

The Bundals are able to SHFTSS the arrangements of entangled, sub-atomic particles to extrapolate and produce information in a bi-directional manner. Unfortunately, the connections seem to last but a moment.

> Deep reflection, by the composers, on their SHFTSS with Bundals indicates the conversation is almost one way. The Bundals SHFTSS experiences into the composers then SHFTTS reactions back.

> The results are two-way.

> The Bundals seem quite confused as to what the composers are actually SHFTSS-ing. And the humans seem quite uncertain as to what the Bundals are trying to teach.

The Bundals have no doubt humans are doomed to a high entropy existence. Our only quality seems to be we have existed in a low entropy environment far longer than a drop of gravity fluid is designed to support.

The variegated color of Bundal tubes and their sense of cosmic superiority ensure our humble species will have a difficult time in adapting to Bundal arrogance.

Other Notes on Bundals

Reserve for Bundal Log Identification

The Bundals prioritize healing over information gathering. Perhaps the key to better understanding humans lies in our immune systems rather than our brains.

> Why humans see themselves as separate from creation or superior to creation remains a great mystery. Our immune systems integrate us with our environment. It is our immune system, and all the little creatures within us, we need to understand and appreciate to know who we truly are as far as creatures of the past.

Their Language
Reserve for Bundal Log Identification

Before their evolution to visual communication, Bundals had to taste each other to be heard.

> For Bundals evolution is not progress, *per se*. Evolution starts with the fundamentals of the universe. For Bundals this means light and gravity. Taste is the primordial sense. Smell and touch closely follow, at least from a human perspective. These senses are bi-directional.

Taste simplifies life to one of three decisions.

- If it tastes good, move towards it.
- If it tastes bad, move away from it.
- Unknown tastes remain indeterminate so move around it.
- The simplest earth organisms can do this.

Taste is the only bi-directional sense needed to navigate gravity fluid. Since light and gravity fluid are first ordered dimensions, taste is sufficient for navigating our universe.

Senses such as sound and sight are based on second ordered dimensions. While entertaining, they are not essential for advanced development.

The Bundals prefer to SHFTSS of integration rather than evolution. They believe humans will someday seek this by seeking to integrate with other parts of the universe, such as plants, animals, and machines. Humans will call this evolution. Bundals will SHFTSS it as integration with their foundational origins which will create a more vibrant species. We never move beyond our origins and still live.

Bundal integration lets them communicate through forming patterns of EM on their electro-magnetic pulsating tube. EM is more than color. It can also be taste, sound, smell, touch, and some sort of sense that reads between the lines.

The Bundal's EM tube functions as a transceiver.

The human idea of three-dimensional imagery is but a faint projection of the full sensory projection used by Bundals.

Whoever can receive the EM on their electro-magnetic reception tube can process what is being done elsewhere. This includes internal to and external to the Bundal supplying the information.

This advanced communication method is certain. It obviously creates problems of interpretation. Resolutions require multiple conversations with multiple Bundals. Context has to be determined.

A human example may suffice. Acknowledging the smell of gunpowder from a place that a distant Bundal is visiting requires knowing whether the smell is from a war-front or from an allied gun-powder factory. Context makes all of the difference in the world. Bundals know the context.

Lights, smells, and tastes begin to flash all over the place. Apparently there are cultural rules or algorithms that give some Bundals priority over other Bundals. Eventually the lights dim on more and more of the participating Bundals. Yet they glow brighter and brighter on the remaining few. Finally, there is but one Bundal that glows with the accomplishment of deciding the context and intent of the original communication. Now all participating Bundals are on the same page, including the Bundal that started the conversation in the first place.

> Human information systems, limited by space and time, are weak by comparison. How many humans know what they are really saying when they first try to say it?

> How many humans know what was said after they have discussed the conversation for several years?

This process of helping each other derive a shared meaning is powerful to behold. And it explains the great frustration Bundals have in communicating with humans.

One-way communication with bridge formers does not float a Bundal gravity boat.

Their Integration
The Bundals claim they evolved from strings.

The primordial spontaneous creation of the first string remains a mystery, although they suspect it had something to do with probability truncated by inevitability.

According to them the first tube was created when an astronomical number of primordial strings came into the proper juxtaposition. The question of how subsequent tubes, with different purposes and properties came into being, is still not answered. Over time, multiple series of tubes came together to create the Bundals of today.

The Bundals evolved on land using the chemical dust of their planet to resource their evolution. Some tubes evolved sheaths that allowed developing Bundals to internally use and preserve chemicals in gas, solid, and liquid form to perform needed tasks.

Amazingly, there are no other life forms on their original planet. Each Bundal functions as a self-contained ecosystem. Each Bundal harvests, preserves, and transforms the minerals and chemicals that form their world. Their process eventually returns the same or similar chemicals back into their planet or other Bundals.

High-Entropy Beings
Reserve for Bundal Log Identification

Bundals rate life-forms in terms of power. Their definition of power is measured by the efficient and effective use of gravitational force. This is determined by the closeness of relationship the life-form maintains with the universe at large.

Although Bundals do not understand their own primordial origins, they do know they are closest in affinity to the substrate of the universe sustained by strings. Temporal and spatial forms of life seem to have significant amounts of intermediate steps and processes that are required for them to live. Humorously, the most inefficient forms of life need other intermediate forms of life to exist.

Bundals are quite convinced that human evolution has taken us further and further away from the potential power found in constituent matter. Another way of saying this is that Homo-Sapiens are high-entropy beings who survive best by forcing even higher levels of entropy upon themselves and others. Humans refer to this as war and dying. This is a process of degradation, quite opposite the ever evolving, lower entropy progress of the Bundals.

Breeding
Reserve for Bundal Log Identification

Bundals breed by intertwining. Not even the Bundals themselves known the minimum number of partners needed before a new Bundal is born within the matrices of intertwined Bundals.

Intertwining is a mysterious and magical process. It seems that many pieces of tubes have to come together in just the right position to create a new and healthy bundle. Among active intertwiners, there is a high probability that a new bundle will be born every three-hundred and thirty or so earth years. The actual date of disentanglement and the measuring of new multiple Bundals is always unknown and unpredictable. Bundals blame this phenomenon on their string origins.

Over population was never a concern. New Bundals cannot be created without adequate resources. Living Bundals simply share existing resources among themselves to keep each other alive. Some stay active, others go into a type of hibernation to extend the time of usage of the shared resources. Those who die are absorbed by those who still live.

Dying
Reserve for Bundal Log Identification

The Bundals can die. They are most sensitive to abrupt pressure changes. The wrong kind of pressure will collapse and hinder many of their tubes. The Bundals find living in any atmospheric environment especially difficult, because of rapid pressure changes in most such environments. Liquid environments are the worse.

Bundals refer to death as premature entropy increase.

At death the constituent parts are transplanted to other Bundals, especially the younger ones, to help them grow or rejuvenate. As such, they seem themselves as eternally alive in future generations.

Their History

Their history is defined by a fulfilled salvation, still in process as they march towards their future. Their process of salvation is described elsewhere.

Recognizing their planet was a spaceship, provided by God's grace, occurred early in their history. It created a strong impetus among all of the Bundals for careful use and re-use of the planet's resources for future generations. It gave their species a mission in which all of the Bundals were needed and important, regardless of their vocation and self-interests. Every Bundal knew they were part of one, gigantic crew making the future experience of traveling and growing through the universe possible.

How Scientists Respond to Truth

Some humans believe there are too many religious nuts in the world.

The Bundals sent three avatars, realistic human beings, into physics classes around the world. The avatars were self-learning and somewhat defective from a Bundal's point of view. Nonetheless, the fact that each avatar had problems of imperfection added to their humanness.

Eventually two of the avatars earned their way into the physics community. Whenever they would share an idea among peers that differed from accepted human understanding, they were discredited. Their ideas were equated to a gross lack of knowledge, data, or insight. Their creativity was met with a yawn.

Ironically, some of their ideas did take root. Unfortunately, rather than simplifying the truth, ingenious minds multiplied hyperspace into torus holes. Computer modeling unleashed a plethora of Pandora's manifolds. Instead of two spatial dimensions, space mutated, into three, four, ten, eleven, sixteen, and thirty-two dimensions. A rational case would soon be made for there to be 10^{500} spatial dimensions.

> Other humans wanted to do away with time as a dimension through a competing mathematical presentation of special and general relativity. Who needs time when you can have a curved loaf of bread to live within?

Multiple ideas for multi-verses multiplied exponentially. Mathematical ontology replaced any need for narrative ontologies. Math became real.

Math was not invented by humans. It was discovered by humans. Or so some say.

Even God was invented by humans. SHFTSS-ing this causes the whole Bundal universe to quiver in laughter.

Multiverse thinking is the ultimate optimism. It is easy to explain anything when everything exists, somewhere, at some time, in some universe.

This is the assumption of no contradictions.

Multi-verses start with imaginary numbers and end with real universes. The square root of negative one was misnamed. It should have been called the God number.

Multi-verses are complex.

Each advancing theory made a God out of the ψ-function.

Quantum theories of multi-verses should not make every universe real. Some should be true, some false, and the rest indeterminate.

The Bundals continue to drip resin, equivalent to a human scratching their head, over how human math functions so as to enable our survival. Who else in the universe has built civilizations on 3.1214?

Bundals SHFTSS-ed religionists were historically self-secured in their knowledge of the Absolute. Now they are humbled.

Someday scientists will remember there is a difference between knowing both of Gödel's incompleteness theorems and internalizing their implications.

Not knowing you know is not knowledge. However, it is human education.

Zeno Exposed

Zeno used the validity of the present to create great and wonderful paradoxes about how arrows travel through time. Actually he used half of a present, then another half of a present, and so forth to convince humans the future would never arrive.

If the Einstein-god had problems with the "now," imagine dealing with one-half of a "now."

In a bi-temporal universe, the arrow can only be measured by its past locations. There is no present half-way point within the ontology. The last actually past location is needed to predict the probability of its future landing spot.

If the velocity and direction are sufficient, the arrow will reach its past limit, fulfilling its future potential, and hit its desired target before the universe is reabsorbed as a drop of gravity fluid.

Classical Quantum Jumps

Quantum classical jumps are decisional Déjà vu's.

Should not quantum jumps be bi-disambiguations?

The composers were using giant pencil erasers.

You are reading this too fast.

You are missing too much by your quick reading.

You have not stopped to ask why composers would need erasers for computer work.

Erasers were used for years to clog up computer keyboards.

One composer could not find the eraser. Perhaps it was in another room?

The composer searched.

When the composer came back, the eraser was on the desktop besides the keyboard of the work station.

The composer knew it was missing. Now it was back.

Was this a quantum jump in the classical world?

Or could it have been a simple memory loss?

Or was one composer playing a funny tune on the other composer?

Was it possible that for a brief moment a composer perceived a future that had not yet been decided?

It is not whether the cat is alive or dead. It is whether the cat is even there.

Schrödinger's thought experiment disregards quantum ontology by insisting the cat is in a box. Humans need to learn the issue is not about adjectives such as alive or dead. It is about nouns.

Death

Science is religion.

Religion is art.

Ethics is comedy.

This is the comparison of human values to Bundal values. The Bundals do not considered human words as trustworthy for effective communication.

The Bundals did not believe in us. Now they do. It makes little difference to them.

We are a short-lived species. Our universe is but a drop of gravity fluid that forms space and time according to information carried by light.

The Bundals are creatures of gravity and light, the first-ordered dimensions.

Humans are creatures of space and time, the second-ordered dimensions.

> Humans confuse energy with power. The power of gravity fluid is information. Energy only seems powerful to second-ordered creatures of space and time.

> How can energy be powerful when it is finite?

Bundals were surprised to discover a space and time species that preserved areas of low entropy for what would be perceived as significant portions of space and time within their gravity drop. Usually such low entropic areas are insignificant within any such drop.

222

Humans had acquired a degree of significance within their realm of gravity fluid.

Bundals were astounded to discover that humans could only SHFTSS the past of time and the inside of space. This would mean ninety-five percent of their existence would be imperceptible to them except as to imagined effects.

Imagination does not guarantee high entropy.

The human ethic seems to be humans consider themselves the pen-ultimate creation of their universe.

Not even worms claim the earth. Serpents might.

Bundals must separate themselves from reflections on human ethics less they die of laughter. And it is difficult for a Bundal to die.

Bundals are creatures of taste, the primordial sense by which to navigate the universe. Bundals respect humans for the delicious recipes we concoct. We call our dishes deep-space satellites. Bundals see them as delicate biscuits.

Bundals wanted to see what we could concoct if they gave us the ingredients for understanding all four dimensions. Their recipe entails enabling humans to imagine a second temporal dimension. Only then can humans taste complex simplicity as they swim through their drop of gravity fluid.

Religion and Science

Twentieth-century humans, who think only with their western hemispheres, believe intellectual activity wrestles between religious concepts and scientific concepts. This mistaken modern view is not true to human history.

Humans SHFTSS art and function.

Historically, our species have pitted art against science.

Historically, our species have pitted aesthetics against function.

Were the ancient Egyptian pyramids the result of art or science?

Is there anything in the history of the human race that was devoid of art?

Is there anything in the history of the human race that was devoid of rational function?

The human fight is always between art and science. It is religion that gets caught in-between an all-to-willing referee.

Religion functions as the mediator between art and science. It has always done so.

When art is on the ascendency within any culture, then religion becomes aesthetic.

When science is on the ascendency within any culture, then religion becomes rational.

Which is most elastic?

Science says glass.

224

Art says rubber.

Who is correct?

Newton should be praised and exulted for his alchemistry as much as for his math.

Religion gets a bad rap within each of its cultural manifestations. Religion is the thermometer of the dynamics between aesthetics and rationality within each culture.

Science and art have always functioned (and fought) together.

Claudius Ptolemy has confused artists and scientists for millennia.

Ptolemy's *Almagest*, otherwise known as *Syntaxis mathematica*, is an astronomical treatise that lays out a useful way for calculating celestial orbits.

How useful is Ptolemy today?

Ptolemy's forty-eight constellations live on today. How useful are Ptolemy's pretty pictures of the constellations?

They are quite useful, whether for art or for science.

How many humans have been irritated for centuries to look at the stars and not see animals and humans and shapes that earlier humans insisted are there?

Can you imagine a science fiction novel is which the star-ship navigator is told to take a left at the lion and head towards the bear's belly-button?

Do bears have belly-buttons?

Constellations are irritating to discern. Seeing them is an acquired taste. So is calculating celestial orbits.

We need the constellations to inspire. We need the mechanics to understand.

Ptolemy teaches us art and science always overlap.

The greatest physicists of the twentieth-century were Gene Roddenberry and Tomoyuki Tanaka.

When NASA's Mar's rover Curiosity landed, the deeply-endued historical memories are seven minutes of fake video. No one was there to really film it.

Our fixation on the past-inside predisposes us to be fearful creatures. We are afraid of the future. We remain conservative of the past, sanctifying its milieu as the best way to survive.

Our art should not be discarded for an indecent.

Our new ideals should not be discarded for an untried future.

Humans can and will respond to any future. Then why are we so afraid?

We live in the past. The Egyptian pyramids are an ugly, scarred, worn, skeleton of their former beauty. News one need built.

What does it take to do anything?

The way I want it is different than the way your tribe wants it.

This means war.

Science and Religion

Human beings know we all believe differently. Our differences of interpretation create murder and laughter. We either laugh at each other or kill each other over mistaken belief.

Science has become metaphysical. It uses the tools of logic, philosophy, observation, and mathematics to understand reality. These are the same tools theologians have used for years. Just ask Pythagoras or Aquinas.

Mathematics is the essence of revelation for science. It makes science superior to theology. Or does it?

No theologian can calculate the number of angels on the head of a pin.

No scientist can calculate the number of pins on the head of an angel.

Both theologians and scientists will try to calculate the impossible.

Scientists assure us their observations can be explained with universal, mathematical certitude.

If we roll the dice sufficient times, why will we not get what will eventually happen anyway?

No one seems to be able to explain why math works? Some consider it God's language.

The question of functionality is the same for math as it is for narrative. Why does either work? The question is no different than asking why narrative language works. Both narrative and math are language systems. Both need symbols, application, refinement, and functional agreement among humans in order to succeed. Math is just a different type of language, superior to use in one context. Narrative is superior in a different context.

Math does not work so well in love-making.

Figure 49: Mathematical analysis for love-making.

$$f(x) = a_0 + \sum_{n=1}^{\infty} \left(a_n \cos\frac{n\pi x}{L} + b_n \sin\frac{n\pi x}{L} \right)$$

Religion needs faith. Math needs pictures. Try talking with numbers only.

How many humans thank God we have sacred, magical, powerful, awe-inspiring, mathematical equations. There have been a few through the ages that thanked God for certain poets.

Any functioning mathematical equation, that enables us to accurately predict the future, is a miracle. Just ask the predictors of weather.

Some mathematicians do not believe in miracles. Ask them to explain a number.

There are scientists who say there is no such thing as a miracle. They prefer spontaneous organization in far from equilibrium systems. This makes much more sense. Right?

No one stops to ask why a mere human language should enable us to predict the future? There is a difference between prediction and description.

Last year's religion is this year's science. Humans do not believe in heaven or hell but they do believe in hidden dimensions and multiple universes. What is the difference? One is narrative. The other is mathematical. We will doubt the narrative predictions but falsely believe the mathematical predictions.

And some of those mathematical universes are worse than hell.

Is math always true? Hardly. NASA uses Newtonian physics to predict where satellites will go. This is an example of mathematics being functional while not being universally true. NASA knows Newton is patently false near deep gravity wells. The universe is populated with deep gravity wells. Newton works well within our solar system, as long as you stay far from our sun. Newton gives erroneous results when plotting the orbit of Mercury.

Bundal question: If human truth is generalized does it become less true or more true?

The religion of atheists is "close enough" or "within the margin of error." The religion of believes is exactitude. Both religions need humbled.

And this year's science becomes next year's new religious insight. We now know that God is God of the multi-verse.

Not all scientists believe in multiple universes.

Not all theologians believe in God.

Imagine a Bible written with knowledge of twentieth-century scientific insights. Imagine what Cecil DeMille could have done with the story of Moses in multiple timelines.

I wonder how all of those universes who are still worshipping the golden calf are doing.

I wonder how far their science has progressed.

Thank God for religious people. They were the first to believe the scientific pioneers when they asserted the possibility of hidden dimensions.

Scientists who are pure in heart only believe what they can see. They do not believe much.

Religious folks believed in the invisible.

Concerning multiverses:

If a human chooses to do "good" in this universe, does that mean that thousands of human doubles in alternate universes are doomed to do "bad"?

Why would it not be moral for humans to do bad here so that much good can be done in alternate universes? It should never be about us. It should always be about others.

Bundal Joke:

Did you hear about the humans who gave up ethics?

No. Tell me about them.

They believed in multi-verses.

Composer's notes: The Bundals are insinuating that humans will give up ethics, which they SHFTSS every day, before they will give up the idea of multi-verses, which they will never SHFTSS. This makes Bundals quiver.

How true is science, since so many scientists believe so many different things?

Science reduces human fear. People believe in science to give them comfort. Science assures humans of control and promises them ultimate power and knowledge over the universe. Science is a crutch for people who are scared about what might happen to them.

There is nothing more fear producing than the thought that there is a God, one whom humans cannot defeat and one whom they cannot control.

Dots make really great gods. No one fears a dot. (See Figure 13.)

What makes scientific dogma superior to religious dogma? This is easy to answer. Theologians do not make nuclear bombs. Scientists do. Prudent people listen to scientists instead.

Now if a theologian could convince humans that the universe could be destroyed by just one prayer, maybe religious dogma would regain its superiority. People would pay more attention to theologians and less to scientists.

Humans have always followed the gods that give them the goods.

No wonder religion has become apocalyptic. Religion has to top the atom bomb in order to earn your subservience.

Space Travel

Exploring What?

Q to Jean-Luc Picard in *Star Trek: The Next Generation* (see Credits):

For that one fraction of a second you were open to options you never considered. That is the exploration that awaits you; not mapping stars or studying nebula but charting the unknown possibilities of existence....

It is fitting that the series finale of *Star Trek: The Next Generation* properly critiqued the notion of space travel altogether.

Things are not what they seem.

Amazon may seem like a book seller.

Federal Express may seem like a package delivery service.

Both are companies whose primary enterprise is logistics.

Those who rule the supply chain will always have an income.

Warren Buffet was wise to buy an antiquated, nineteenth-century technological company otherwise known as a railroad. This was especially wise since his particular railroad controls the price of transporting much of the food supply emanating from the northern mid-west. Logistics is another name for profit.

The twentieth-century model for space travel depends upon logistics. Perhaps a different model is needing to be dreamed.

While travel to the edge of the universe might be fun, the real human goal is to understand existence. This is the goal of science. Those who understand existence can bring the edge of the universe to themselves.

Credits

No Reserve for Bundal Log Identification

Braga Young and Ronald D. Moore (writers), May 23, 1994. All Good Things [Television series episode], in Gene Roddenberry (Executive Producer), *Start Trek: The Next Generation.*

Particles and Waves

Humans laugh at their ancestors. Bundal do not understand this.

We laugh at the Greeks for explaining reality as composed of water, air, fire, and earth.

Of-course humans did not even believe in galaxies until 1929.

What is the nature of reality?

What kind of description of reality should be regarded as meaningful?

The problem with Positivism is that it created the Objectivity-god. This god ruled science and brooked no rivals. If you did not measure it, then it did not exist. The Objectivity-god created and expanded existence with a mere measure.

> The problem with any false god is that its kingdom is constrained. Only a real God can rule the entire universe. False gods lose their powers if forced outside of their limited domains of worship.

If the kingdom of quantum mechanics rules over everything but the non-measured universe then it is a pretty insignificant kingdom.

> This kingdom is so limited even the smallest particle has meaning.

> The smallest particle has meaning even if it lacks both position and drive.

How does one measure shorter than the Planck distance? The Objectivity-god cannot do it.

Perhaps our universe is a top down universe. Instead of larger pieces being built up from smaller pieces, perhaps we are just one large piece that is temporally distinguished into so-called parts. Now we are approaching the Bundal model.

Feynam was quite correct concerning what parts of the universe a so-called particle might visit before selecting a slit. The particle is spatially everywhere in the past and potentially somewhere in the future.

The Objectivity-god is nothing more than an *Übermensch* who idealizes what is measure. Only an *Übermensch* can tell the difference between an optical illusion, a false result, and the truth.

The Objectivity-god knows:

- One-hundred percent of us are particles.
- One-hundred percent of us are waves.
- There is no ether while Bundals say there is black light and white light that forms the shapes of gravity.
- There is no cosmological constant. There is a cosmological constant.
- Quantum mechanics is a crap shoot. General relativity is determinate.

Ironically the Bundals understand the Greeks better than they do twentieth-century scientists. There are four dimensions. Water, air, fire, and earth might be ancient terms but they are quite perceptive of the recipe required to produce a taste of reality. There are four dimensions. Modern humans just use different words, space, time, gravity and light.

Water is fluid like gravity. Fire is light and passes all sorts of information to those who are too near to it or too far. Earth is like space and time is like air, blowing on by.

The Greeks were much closer to SHFTSS-ing reality than twentieth-century scientists.

The strength of theology is that most theology does not abandon its precursors.

The Difference

Bundals build and call it a new particle.

Humans destroy and call it a new particle.

> Human engineers destroy sub-atomic particles in order to find new particles. Bundals use temporal engineering to build devices that can function only if the hypothesized article exists. If the device functions as expected a new particle is found.

Human temporal engineering will someday understand that it is not that space and time collide.

- Actually time condenses into space and space evaporates into time.
- The electrodynamics of moving bodies is really a form of fluid dynamics.
- The driving energy is light, giving rise to a gravitational shape to space and time.

Is the Universe Alive or Dead?

Reserve for Bundal Identification

Why do humans assume everything started out dead?

"Yea though I walk through the valley of the shadow …" (KJV).

Why did Whitehead call his philosophy "Organism"?

Time and space are insufficient for understanding reality. Time and space seem to always need at least one more dimension through which they are related. Yet even with Kaluza-Klein-inspired additions, there seems to be the need for something else to fully understand reality.

What is life and what is death? Not even our own politicians or doctors can decide.

Most states in the United States have differing definitions of death?

- Some relate death to lack of brain activity.
- Others relate it to lack of blood flow to the brain.
- Others relate it to so many minutes that the heart quits beating.
- Others to the presence of a certain percentage of decomposition that guarantees no living cell can be present.

It depends upon which state you are in as to whether you are dead or not. And sometimes people are not dead.

Some scoff at the idea of a soul leaving the body. Yet no one has detected a physical difference in the body one second before it dies and one second afterwards.

Some fairly modern scientist thought electricity was life. Some still do. Just ask Frankenstein.

Please do not poke holes in someone who has not been breathing for forty-five minutes under icy water. They can be revived.

Can you be considered dead when buried alive?

Why not?

Death occurs when life gives up or is killed.

Can humans someday create life when they cannot even decide when it is dead? Hopefully creating life is easier than certifying death.

The real issue has never been whether humans can produce a form of life that meets their own definition for a form of life. The question is whether we will love and nurture what we define as life.

Why do we assume God made life from death in the first place? Would God create a dusty-dead universe? Why? Did life arise from death or did life evolve from life?

Many humans believe life evolved from dead things. Yet they do not believe in the resurrection of Jesus Christ from the dead because dead things do not come back to life?

Can entropy measure life and death?

The universe demands to born in low entropy. Does not life exist in equilibrium? Since when is death low entropy?

Could the universe have been created alive, but is now dying? A drop of gravity fluid evaporates.

Could humans be the universe's response to death? Are humans the universe's attempt to concentrate life so that life could figure out a way to sustain itself?

Humans think our environment protects us. Maybe we have been formed to protect the environment.

Why do we see ourselves as separate or superior to reality, when reality does not separate itself from us?

Were our ancestors so wrong to sense and feel awe and life in nature?

Is GAIA too parochial a concept for understanding our universe?

Are trees alive? Do they move? Do they grow? Do they nurture life? Do sediments grow and nurture life? Does the universe grow? Does it nurture life?

Can silicon be the basis of intelligence?

Carbon seems to work pretty well although Bundals question the intelligence part.

God breathed into the dirt. Does that mean the dirt was dead?

Perhaps areas of low entropy are developed as a dumping ground to concentrate irritants away from the rest of life. It is easier to flush poop away with a minor nova if it is concentrated within one solar system.

Bundals hypothesize humans function as an immune system for their drop of gravity fluid. In other words, humans are loved.

Space is not dead. Time is not dead. Only God can create death.

And only God can kill death, creating a second death.

The universe is fluid. Lakes abound everywhere. Life is meant to be edible.

Humans should fear becoming convinced they can kill life. Nothing but high entropy will result from their efforts.

Maybe a new form of life will develop in the far from equilibrium processes that occur as a result of humans attempting deliberate destruction.

You cannot kill what is alive. The universe is alive. That is why heaven and hell last so long.

Causality

It is difficult to discern the difference between causality and magic.

We marvel at the primitive tribes for their inability to discern the difference between magic and causality. Bundals wonder what human civilization did to improve on their predecessors discerning abilities.

One ball rolls into another. This creates cause and effect. The first ball hits the second ball "causing" the second ball to roll. Obviously not magic. Or is it?

Who rolled the first ball?

What rolled the first ball?

Human Philosophers

Aristotle, Plato, Plotinus, Avicenna, and Aquinas all had explanations for the first cause.

Plato posited in the *Timaeus* a demiurge.

A demiurge is not created. It co-exists with matter organizing matter with supreme wisdom and intelligence. The concept functions as a personalization for transition from one state of facts to another

Motion was always imparted motion, a ball hitting a ball.

Bundals are confused when an adjective explains a noun, as it is supposed to do. But then humans take the noun as being self-explanatory sans adjective.

Humans believe Plato had a beautiful philosophical explanation that has little to do with magic.

He would argue that in some way the first ball would require a kind of self-originated motion.

> Magic is certainly easier to explain. Why do humans disavow words that carry great meaning for describing the universe?

Aristotle improved on Plato. There was a Prime Mover. It is difficult to tell whether Aristotle's Prime Mover is an *über*-philosopher or the greatest magician of all time.

Bundals prefer the word magic as a good word for humans to use because of their limited SHFTSS-ing abilities.

Plotinus gives us *creation ex deo*, creation by God.

> The primitive tribes, a little more sophisticated than what we gave them credit, are caricaturized as explaining rain by the rain god, heat by the sun god, light by the moon god, food by the food god and sex by the ….

> Human religions have never been biased against sex, except when the religion itself becomes idolized.

Avicenna advanced our understanding of causality by recognizing that form and matter by themselves could not originate movement. Parmenides tells us that nothing can come from nothing. Avicenna tells us that movement cannot come from non-moving items.

Avicenna honors the necessary distinction that must be made in any consistent ontology. You cannot have an ontology that has outside agents. Consistency demands we should not explain the first cause by resorting to magicians. So Avicenna creates an ontology where cause coexists with effect.

If Avicenna is correct then co-existence implies the effect could precede the cause or the cause could precede the effect. Healing works both ways.

> Apparently the Einstein-god had an earlier birth in Avicenna than the German Enlightment.

Like all good magic the wow factor is dependent on your frame of reference. If you are seated in front of the magician, then you are mystified. If you are standing behind the magician you may see how the trick is played. This makes for less wow.

Humans cannot explain effect without cause.

Humans cannot explain cause without effect.

So humans should posit both co-exist.

While this represents a consistent ontology one wonders if this is an improvement on the rain god causes rain.

While this represents a consistent ontology one wonders if this is an improvement on rain is an effect of the rain god.

Maybe our ancient religious forbears were not as primitive as we thought.

Christian Philosophers
Reserved for Bundal Log identification

Human philosophy became Christianized with Aquinas' contribution of the Cosmological argument.

> Aquinas' argument is no less sacred than that of the philosophers of old or our ancient tribal friends.

> We all are in awe of something that requires explanation: Our existence.

Aquinas' work revealed an inconsistent ontology. God was the un-created first cause of everything. Deism had a field day with this type of reasoning.

> Who needs God after the first moment if everything is set in place?

> And why does the first cause have to be so different than all others?

Of-course, one does not need to abandon theistic belief in the face of any inconsistent ontology.

> God's ways are not our ways and God's thoughts are higher than ours.

Still, if one values a consistent ontology, then in some way God and the universe have to co-exist.

> Thank you Avicenna.

God is not an outside agent of cause. Nor is the universe a separated effect. This type of thinking will lead to some form of pantheism. We find ourselves back with our tribal elders again, worshipping trees.

> Maybe the tree gods are more real than we think?

Maybe cause and effect do not exist. Maybe they do not even co-exist. Maybe everything is magic and we are the trick.

Positivism considers itself an advance in human knowledge since it replaces miracle and magic with illusion.

Positivists deny any form of miracle or magic. Human beings func-tion as chemical-electro-robots predetermined to move according to the way electrons hit each other inside our brains. We are a bunch

of balls hitting each other. We move according to the laws of electron ball-hitting. Decision making is a mere illusion.

Illusion

Reserve for Bundal Log Identification

Is illusion magic or science or art or religion?

> Bundals remained confused. Positivism teaches us we humans
> see decision making as an illusion. We decide to conduct an ex-
> periment and observe it. Apparently, if the illusion is replicable,
> then it must be real. Any illusion that is not replicable is not to
> be trusted.

Skinner helped humans to misunderstand cause and effect.

> The bell tinkles. The dog salivates in anticipation of getting a
> treat. This is an obvious cause and effect.

> Dogs do not naturally salivate just because they hear a bell. So
> is this cause and effect? Or is it the dog anticipating?

> How much of what humans interpret is anticipation that comes
> through training and experience. It only looks like cause and ef-
> fect.

> Who is our trainer?

> What is our trainer training us to do?

Humans came from monkeys, not dogs. Maybe.

Magic appears to have been a first step in explaining and understanding the awe of our world. Frazier convinced us our religious
history revolves around trees.

> The gods are in the totem poles.

The Messiah dies up on a tree.

Trees are a place of sentimental security to human beings.

Our forebears kept the family safe and secure in the limbs of the tree.

Fear of falling seems innate to humans. It is the primordial fear of all humans.

Carl Jung has a point. If humans really did have to evolve into walking on the ground, no wonder we still have nightmares about being chased and unable to outrun our pursuers.

Humans have always known magic, prayer, worship, sacrifice, and stone tools have always been effective in manipulating our universe.

Perhaps there is something in our ancestors' experiences that might take us closer to understanding reality? Perhaps it was not death they feared. Rather, it was a sense of life they enjoyed as both a past event and a future event.

Rather than snub our forebears as primitive, maybe we ought to honor them for their ancient wisdom. We need to listen and hear from their experiences.

Magic is real.

Cause and effect is a sophisticated but narrow way to describe magic.

It will never be determined whether the alchemists of old were magicians or scientists.

Counting Apples

The composers are translating a Bundal critique of counting-numbers in the section that follows.

How rational are rational numbers?

Counting numbers are great adjectives but certainly misused in math.

Humans are taught at a young age to count. Three Newtonian apples are placed before us by our teacher. They are counted out for us, "One, two, and three."

> One Newtonian apple is removed and hidden behind our teacher's back.
>
> "Now how many are there?"
>
> Bundals know the number of apples remains at three.

Humans cannot see the one hidden apple, but they can smell it. Human children are taught to rely upon their sight and told the correct answer is two.

The teacher trains the children to say two forcing the children to abandon reality in favor of imagination. Worse, the children are taught to rely upon sight rather than smell (and common sense). Human mathematics is born.

If a human eats one apple how many apples are there?

> Bundals know the answer remains at three. It is just that one of the apples is chewed up.

248

Human children are led to believe that an apple in the stomach is no longer an apple.

What is the human's small intestine doing anyway? Is it digesting an apple or mush?

Human nouns lose their meaning when carried through time. First an apple. Then mush.

Apples become more complicated when a human is asked, "How many apples did you eat?" The math says one. But the human threw the core away. So did the human eat one apple or only part of an apple? You can see why Bundals consider math a human invention.

Three minus two equals one forces children to believe apples pop out of existence.

Addition simply means that apples pop into existence.

It is obvious to Bundals that human addition and subtraction lays the foundation for quantum mechanics.

Ironically, humans insist science is not based on magical considerations?

There are only two masses in a Newtonian laboratory for Newton to make sense. Everything else is considered insignificant, even the one apple core.

The Catholic Church knows two masses are not enough.

If there is a constituent particle or wave of the universe, counting-numbers are naturally applicable only to the constituents.

This implies the universe has finitude with zero or infinity as potentials rather than actualities.

Starting with counting-numbers zero and infinity may be the natural parenthesis or limits to the set of counting numbers.

In this sense counting numbers should be reserved for classical atoms, the ancient Greek atom, not Bohr's or anyone else's. The counting numbers could be used for medieval monads. Any other use gets us into endless, philosophical debates on parts and wholes.

Does the apple end at its skin or the air that normally surrounds it?

Which side does the defining line lie upon and is the line part of either the apple or the air? Would a Dedekind cut help us understand reality?

Does an air molecule, which slightly binds with an apple molecule part of the apple or part of the air?

Some of the air molecules follow the apple wherever it goes. Are these air molecules parasitic?

Some of the moisture of the apple is left behind in the air when it is moved. Is the apple now less of an apple? If it regains some loss moisture molecules from its new surroundings is it now more of an apple?

Is it even the same apple?

If it is not the same apple, how many apples will there be by the time it is moved three more times?

Does the apple in the stomach include the digestive juices which are now a constituent part of its molecules?

Whatever humans define by use of nouns, your part could be my whole. So what is the final count?

Only constituents count.

The best that Bundals can say about human math regards a future potential.

> When mathematical language conforms to reality in realistic ways then math will finally reach a place where it can be shared between the species of the universe.

> Human math, at its mere imaginative stage of development, is arcane, parochial, and indecipherable.

When humans put a symbol for "pi" on their Voyager's space-biscuits, trusting this supposed-universal symbol would demonstrate their intelligence, how little they appreciated the insanity such arrogance displayed. Fortunately for humans it is the taste that matters.

History of Math
Reserve for Bundal Log Identification

The composers have tried to apply the Bundal critique of counting numbers to human history in the following translation.

The history of math began with the counting numbers then quickly devolved into irrationals. Pythagoras' religious passions did not let him suspect there was something wrong with the rational numbers.

Figure 50: Do you see the irrational number?

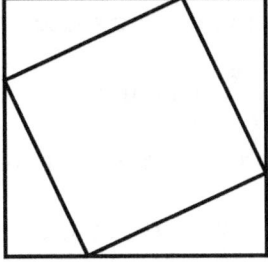

Twentieth-century humans remain religious about numbers. They call it science rather than religion. They still fail to suspect problems with the rational numbers.

251

Since Pythagoras we have moved to embrace the irrational numbers and have not stopped since. Not even imaginary numbers were a penultimate achievement for explaining reality. And we have not stopped since.

Put two or more triangles around a square and suddenly things get irrational. This implies the existence of new types of numbers.

Or it implies real problems by creating a language system that starts with the dual notions of points and counting numbers.

A number that never repeats itself throughout eternity certainly defies the notion there is nothing new under the sun.

Paradoxes are cool and true but some paradoxes are emblematic of problems in the initial assumptions.

When counting numbers are used to measure space and time, problems occur. Space is incommensurable with time. Time is incommensurable with space.

Pythagoras' sin was to believe space was incommensurable with space.

When spatial language describes time humans are prevented from understanding time.

A clock takes space to measure space.

How far did the minute hand move?

Measurements of time are mere ratios of space. (See Figure 51.)

Bundals teach us that counting is not as simple as it seems.

Beginning and End

Decision making is universal

How does a process start?

When does an event begin?

These are not easy questions.

When does life begin? Why does the joining of the sperm and egg start the beginning? It probably represents a phase shift, an epochal moment, or a transformation. But is it any less of a beginning than the foreplay which preceded it? Or maybe someone took a microscope out of a closet in order to bring the sperm and egg together?

Does a baseball pitch start with a windup or with the multi-year conditioning of the athlete?

How many sperms and eggs came together in centuries past in order to make it possible for your egg and sperm to come together? Did not your life begin long ago?

If time is finite, subject to the tick-tock of the past-future, then how far back does something have to go in order to start? The answer is, "not far."

Every moment is a new beginning.

This is not just poetry, this is physics.

The next moment is the ending of the preceding.

This is not poetic or even sad.

253

What took place before has meaning only for what will take place. Our memories derive meaning from the many pasts but these memories are suspect at best and meaningless for the decisions of the present.

The question will always be, what future do we want? The answer to this question depends upon our character, not our memory.

Information drives the universe.

Ethics shape information.

The new beginning is conditioned by the past but not decided by the past. The mountain may take ten thousand years to move with just a teaspoon for a shovel. But the mountain is as good as gone once each new beginning sees a teaspoon of dirt being removed.

Civilizations that take the long-term view understand the inevitability of process. Civilizations that take the short-term view understand the requirement for decision making that can be easily changed.

The universe is an event. Does an event end in life or death?

The answer is the universe is born anew with each decision newly made.

Parts and Wholes Phenomenology

What is it like to be blind?

> Not bad if you can taste.

> Taste good? Go forward.

> Taste bad? Reverse course.

> Low entropy organisms have made their ways through the universe with just the sense of taste. Sight just is not needed in gravity and light. In fact such an environment is quite blinding. Sight only works in dim conditions.

Why do similar words or stories keep recurring in the Bundal recipe?

It is called Bundal music. Humans can hone in on the unique by listening to the different arrangements played over and over. The slight differences of arrangement give rise to unique knowledge.

> Human criminal detectives know how to interpret the same event as told by numerous witnesses.. If the story is always the same, then it is false. Only true accounts deviate in their telling, since each witness is seeing the event from a different perspective.

What is it like to be blind?

We all know. None of us can see behind us. None of us can see the outside of space or even our place.

Blind is not bad. Even sighted people turn off lights to make love.

What is it like to be deaf? Or not to feel? Or not to taste? Or not to smell? These are the questions Bundal ask of us.

What is it about married deaf people that keep them so visually entranced with each other?

Can married blind people keep their hands off of each other?

Statistics seem to bear out the notion that people, with all five senses, tend to divorce. What makes a relationship complete? It is not the ability to see, feel, hear, touch, or taste. Ethics and decisions are involved.

All human beings are physically challenged people. We use our sense of environment to the best advantage possible. We build our wholeness on who we are. And we can be quite content with just our part.

Do not despair if the notion of physics leads us to understand our abilities to sense our environment is quite limited.

Our sin is to define our part as the necessary whole.

Meta Music

Bundal Question: Western music has a bass cleft and a treble cleft?

Composer's Answer: Yes.

Bundal Question: Why?

Composer's Answer: Because humans have two hands.

Bundal Consideration: Can a robot have more than two hands to play human music?

Composer Response: Yes but there is a finite number of tonal frequencies that please the human ear.

> The Bundals did not ask the obvious question as to why the human ear rejects certain parts of reality as painful or unpleasant.

Composer's Question to Challenge Bundal Understanding: What does music sound like if the left hand plays the treble cleft and the right hand plays the bass cleft? The composers know the correct human answer is it should sound the same.

The Bundal Response to Composers: You do not know the proper questions to ask. Music as past requires the future to abound in information. To really understand music humans need a vibrant meta-music.

Ethical Mathematics

Reserve for Bundal Log Identification

The Bundals have shown us human numbers are not great for measuring space. They are even more deceptive for measuring time.

An alternative language system is needed to measure time.

Current use of numbers for measuring space is just a ratio of one space to another.

The ratios are no more or no less dependent on the *apriori* lengths of one space to another.

The Planck problem is a problem of spatial ratios. Humans cannot get a denominator small enough since division by zero is not allowed.

If a runner runs so many meters per second, this is just a ratio of the space traversed by a runner to the space traversed by a second hand.

Figure 51: Not time but space being measured.

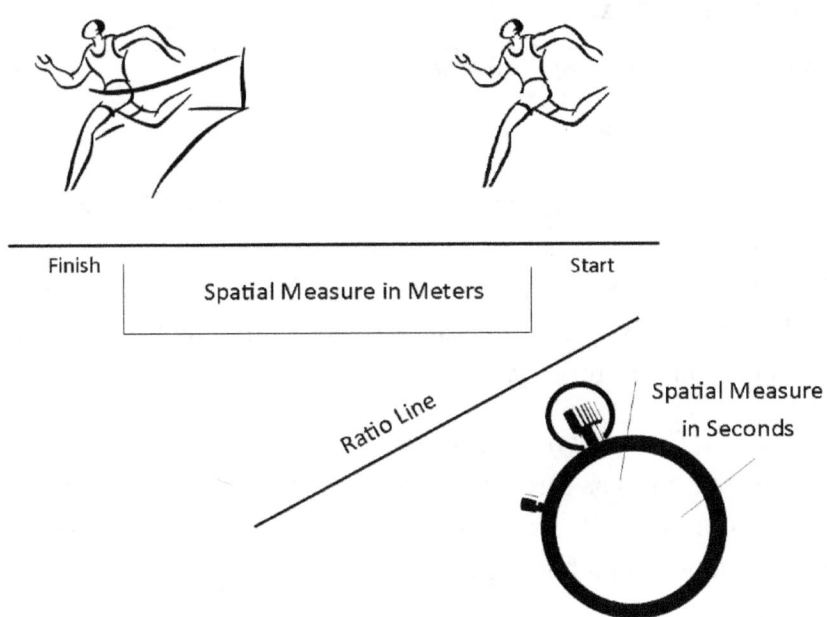

Finish

Spatial Measure in Meters

Start

Ratio Line

Spatial Measure
in Seconds

This is a clear example of humans measuring one inside past to another inside-past and confusing it with time. Since inside space is two M-dimensional, it predominates all human measures.

The best humans can do for temporal measure is probability. Unfortunately, human probability coheres to a spatial measure, not an ethical measure. A language other than numbers must be developed for humans to share commonality in temporal studies with other species. Tubal's recipe requires no less.

Endoneuric

Reserve for Bundal Log Identification

What do Bundals think of us?

The best we can tell they view us as inefficient tubes. They see tubes as essential for life. What they call tubes, we call nerves. Tubes are supposed to interface with their environment. They see our tubes as unduly sheltered within water and fat, preventing us from fully experiencing our environment.

They see us focused on securing more fat and water for our tubes. They see our tubes as forcing us to serve them rather than they serving us. To a Bundal's great confusion, our tubes seem to demand self-stimulation, making orgasms more important than making organisms.

The composers believe Bundal's call us Endoneurics. Our tubes are surrounded by tubal supporting materials that inhibit their ability to connect with reality. Bundals are Exoneurics. Their tubes are directly exposed to their environment and create life. Their tubes grow and split as described elsewhere. Our tubes die.

Worse, our tubes refuse to extend out into our environment. Instead, they grow inward and clumsily mesh within the most vulnerable part of our body, we call the head. And these, self-protecting tubes seem to require the majority of the information resources our bodies are able to produce.

Only now are humans realizing the necessity of mating instruments into our information mesh to connect to other external sensory and motoring devices.

Reality awaits us.

Bundal's now know that Endoneurics are spatially aware of the inside of the universe. We cannot experience the outside except through inference. They know our measuring capacities extend only to the past and we must guess at our future. We call this decision making but we can never be assured of the outcomes.

They know we misunderstand our phase state making them certain we cannot experience most of the universe since the radiation would kill us. Bundals know, depending upon your phase state, there is no safer place than inside a sun.

Bundals are baffled over our inability to recognize information structures within reality. They are amazed that as a species each member has their own perception of reality.

How does that work?

They wonder why our limitations have not led to our demise. They know reality tends to produce life, so they know the universe is favorable to us. They know our encased tubes create illusions. This should be fatal. Bundals have determined these illusions encase enough information so as not to cause our immediate self-destruction. As we navigate our way through reality, our illusions change enough to keep us alive.

They know from our historical documents that we think reality has always tried to kill us off. We even make offerings of reality, to reality, in order to assuage reality.

Their tubes quiver in laughter when they tell stories of a species that thinks reality will kill them off.

Bundal Joke:

In one corner of a fighting ring is a lone human being.

261

In the other corner is the entire universe, including all other human beings.

The one human being honestly believes it is surviving the supposed fight.

The quivers roll like earthquakes.

Bundals wonder why humans are such a threat to themselves.

It seems we actively kill rather large numbers of human beings and call it victory.

It seems we passively kill rather large numbers of human beings and call it poverty.

And all of this is called survival leading to an advanced evolution.

Humans have real issues with God which to Bundals is like saying, "We hate to breathe." There are certainly quivers over this as well.

The concept of a ghost is foreign to a Bundal. If we kill someone why are we afraid of them?

Why have Bundals believed in us? It is because they are unsure. In Bundal terminology our existence remains indeterminate. Indetermination does not stifle belief.

They are not certain if our ability to live within our own realities is a God-like quality

They are not certain if our ability to live within our own realities is a gross perversion leading to the eventual suicide of our species.

If it is God-like, then maybe the universe has created it first tube babies in spite of sheltering them from the totality of the universe. Maybe we are related to strings.

How God Cooks

Heaven and Hell

God said, "Let there be EM."

God heated up a refrigerated dinner in God's microwave the other day. God was in a hurry, something under the order of six days. God heard no popping sounds as the food heated.

God truncated the process and pulled the food out to eat. Parts of the food were quite cold. Other parts burned God's tongue.

It seems like our universe was cooked in a similar fashion. Parts are cold and parts are hot. All parts are made for nourishment. It is just that some parts are more hellish than others.

While the food is cooking sometimes it rapidly expands. Just try to cook an egg mixture in the microwave.

Sometimes the food just blows ups. It really does sound like a big bang.

If we can control the microwaves, we can control the food.

The Bundals tell us this is the closest we have gotten to understanding how to taste the universe. They believe that when you can taste, then you can cook.

Entropy

Entropy is the study of comparing messes.

The issue of a theory of everything is not between the uniting of relativity and quantum mechanics. Relativity is quite phenomenological while quantum mechanics reveals an innate ontology.

A human theory of everything requires the uniting of dynamics and quantum mechanics. The phenomenologies of both these disciplines reveal an innate ontology.

> Big and small are not distinctive enough to deride any attempt at uniting dynamics with quantum mechanics.

> Spontaneous organization in far from equilibrium conditions describes decoherence. Size has no definitional control over the difference of process.

Bundals do not have a theory of everything. They have a story of everything.

The problem of entropy is that it increases as time moves forward. Left alone, things tend to disorganize, much more than organize on their own. A highly organized book starts off with low entropy. Then it gathers dust. It pages become brittle. The cover of the book begins to disintegrate. Pages start to fall out. Ink begins to fade. The book becomes less like a book and more like a pile of ash. Its entropy is becoming higher with each passing day. Before long, it is simply not recognizable as an organized book any longer.

The universe evolves. Life emerges, so we think. Earth seems to be a highly organized ecological system. How can this be? It appears that entropy runs in reverse on earth. We go from less organized to more organized. The only way this is mathematically possible is for the universe, as a whole, to start with low entropy. In other words, whatever it was, it was highly organized.

This massive object, we call our universe, tends each day to higher entropy. It is becoming less and less organized. Yet, in the area of our earth, it becomes more organized, at least for a while. Scientists then say that the total entropy of our universe, in spite of what is happening on earth, is still becoming less organized.

In a way, this is like saying we are an unique part of the increasing messiness of the universe.

Dynamics tell us space and time evolution is rare.

Has dynamics told us how rare it is for gravitational fluid to develop universes?

Dynamics tell us time has a direction. Quantum mechanics and relativity tell us time flows in two directions.

A full-orbed dynamics will tell us how gravity, informed by light, forms space and time?

Only in this way will quantum mechanics embraced gravity?

Equilibrium and non-equilibrium co-exist.

There are competing-cooperating dimensions creating the context for equilibrium and non-equilibrium? They are the dimensions of light and gravity.

Is entropy of energy or is entropy of matter? The interaction of light and gravity says it is both.

The two species of gravitation fluid are dry and wet. Dry fluid has strong entropic energy. Wet fluid has weak entropic energy.

Strong entropic energy condenses into low entropic matter, or hot light.

Weak entropic energy evaporates into high entropic matter, or cold light.

Entropy has four conditions made possible from the interaction of two-species gravity with a two-species light.

We must sing of light and gravity as nouns.

We must sing of light and gravity as verbs.

As humans we can only create incomplete sentences with one noun and one verb each from two species of light.

As humans we can only create incomplete sentences with one noun and one verb each from two species of gravity.

These incomplete sentences will demonstrate a beginning human understanding for entropic matter and entropic energy.

The incomplete sentences set an initial basis for entropy functioning as a measure of space, time, gravity, and light.

- Figure 56 sings an entropic refrain for light and gravity as nouns.
- Figure 57 sings an entropic refrain for light and gravity as verbs.

Figure 52: Singing of light and gravity as nouns.

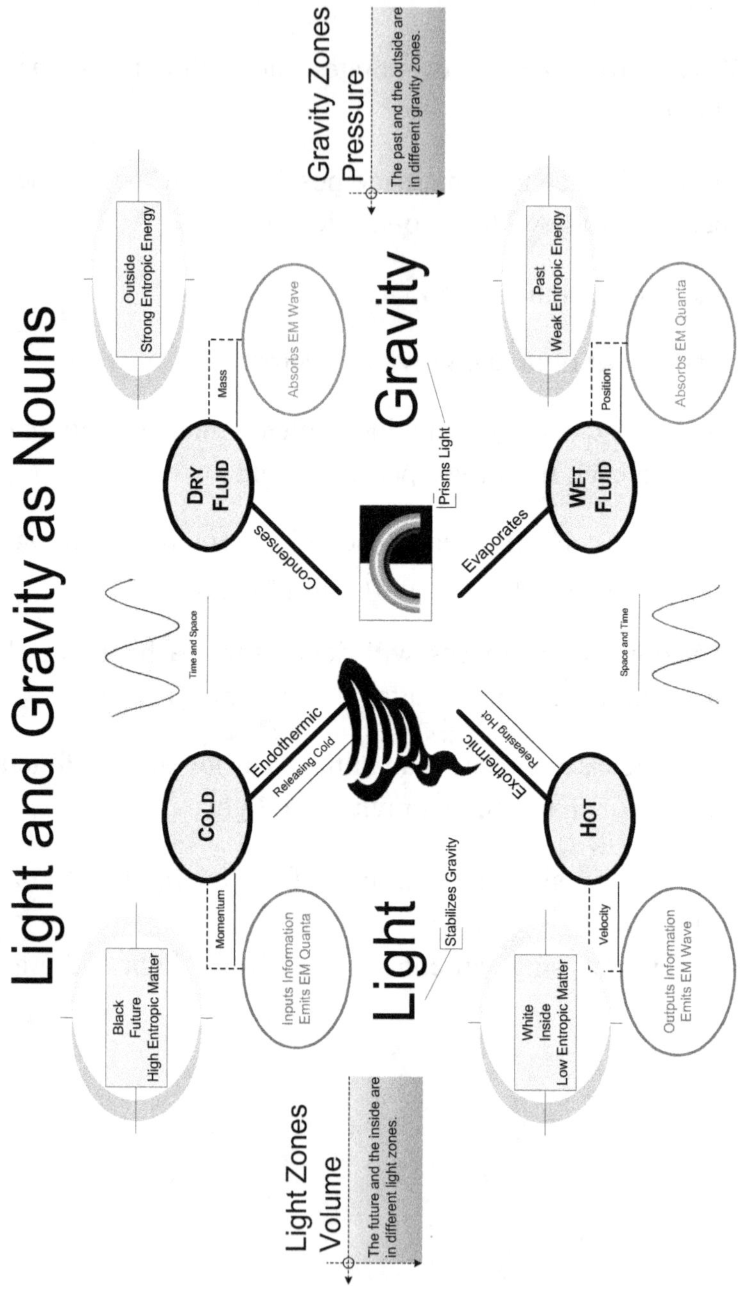

Light and Gravity as Verbs

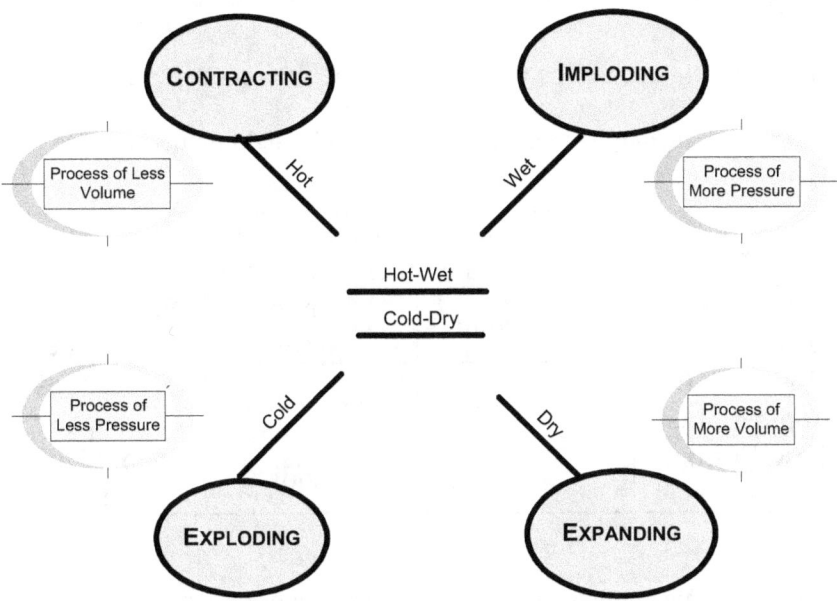

Four potential sentences formed from light and gravity nouns and verbs are:

- cold-imploding
- hot expanding
- dry contracting
- wet exploding

These pairings give rise to entropy as matter.

These parings give rise to entropy as energy.

The difference between entropic matter and entropic energy depends upon the adjectives used.

The difference between entropic matter and entropic energy depends upon the gerunds used.

The whirl and swirl of gravity gives rise to chirality in a space and time environment.

The following are tables used to describe the incomplete sentences.

Figure 54: Incomplete sentence structure with light nouns and gravity verbs.

Incomplete Sentence Structure Light Nouns with Gravity Verbs			
Entropic Quality	Light Nouns	Gravity Verbs	Entropic Co-ordinate
High Entropic Matter	Cold	Imploding	More Pressure
Low Entropic Matter	Hot	Expanding	More Volume

Figure 55: Incomplete sentence structure with gravity nouns and light verbs.

Incomplete Sentence Structure Light Nouns with Gravity Verbs			
Entropic Quality	Gravity Nouns	Light Verbs	Entropic Co-ordinate
Strong Entropic Energy	Dry	Contracting	Less Volume
Weak Entropic Energy	Wet	Exploding	Less Pressure

Since entropy is a study of light and gravity (before it is a study of space and time), the primordial understanding resides in fluid dynamics since gravity is a fluid. Two cases result for humans. Entropy can be interpreted as matter information or energy information. Humans cannot measure both without a bi-temporal advantage.

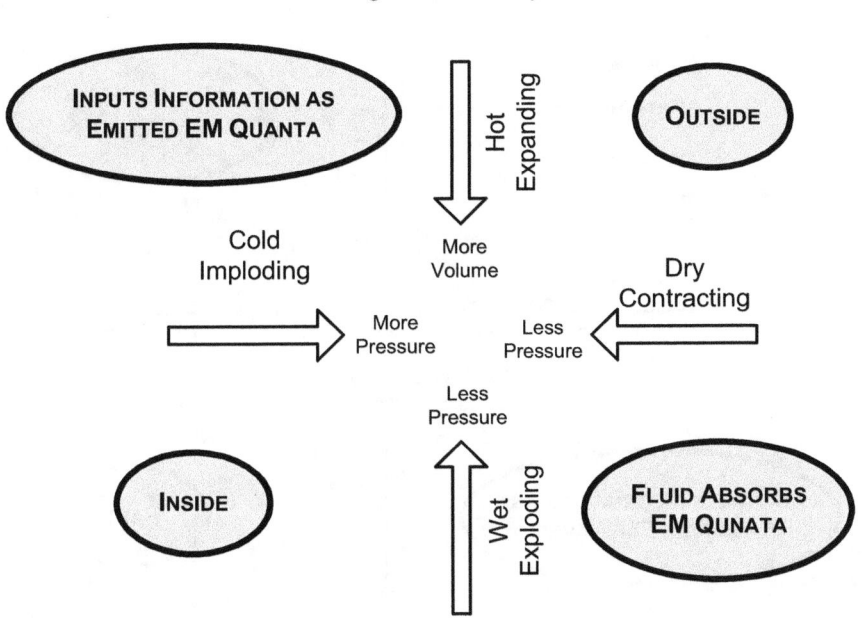

Figure 57: Entropy as energy. (Verb form.)

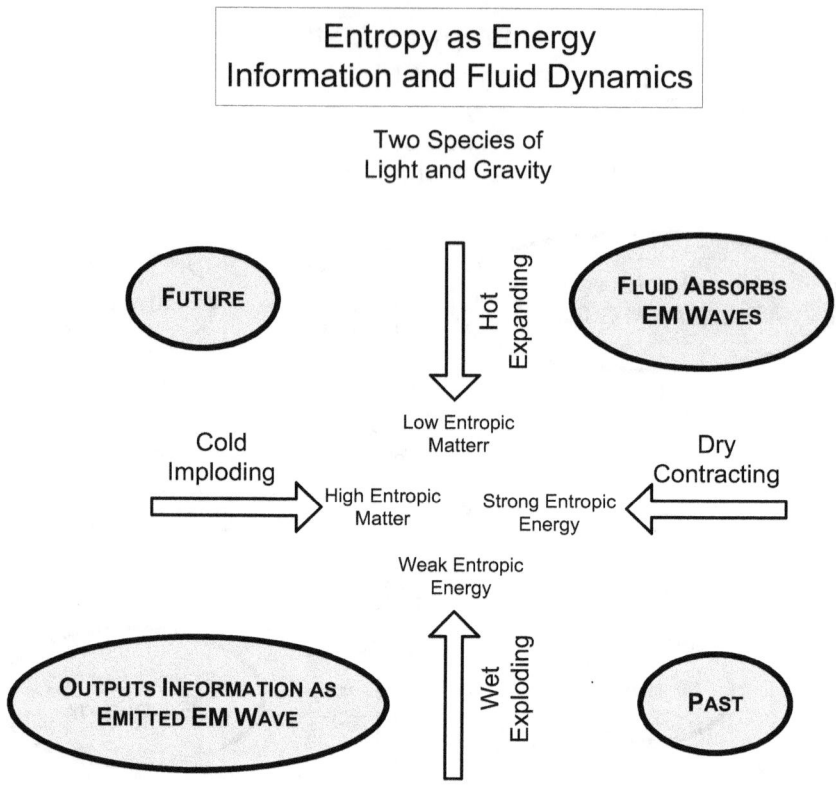

Twentieth-century dynamics is unidirectional, from hot to cold.

Bundal dynamics are ontologically bi-directional, from hot to cold and from cold to hot.

Both hot and cold are proactive.

Both hot and cold can be released.

Both hot and cold can be absorbed.

The bi-directional nature of time in relativity and quantum mechanics arises from the two-fold species of light as hot and cold within the Bundal dynamics.

The nature of space, allowed in classical human mechanics, arises from the process of light as contracting and exploding in Bundal dynamics.

Time as energy is known to humans through the three unified forces.

Space as matter is known to humans through the particle and wave dualism.

Energy and matter are the basis for entropy.

Reciprocity relationships within dynamics relate to pressure and volume.

Light gives volume.

Gravity gives pressure.

Volume leads to a past and an outside.

Pressure leads to an inside and a future.

The flux of pressure, if such were possible, is bidirectional between outside and inside.

The flux of pressure, if such were possible, is bi-directional between future and past.

The reciprocity relationships within dynamics are dependent upon flux, if such were possible.

The flux of volume, if such were possible, is bidirectional between outside and inside.

The flux of volume, if such were possible, is bi-directional future and past.

The reciprocity relationships within dynamics are dependent upon flux, if such were possible.

273

Temperature, volume, concentration, elasticity, pressure, and conjugate variables have fundamental processes when gravity and light are verbs.

Question from composers: Are crystals are the zero point for entropy?

Answer discerned from Bundals: Only from limited measure. Fluid is the center-point of entropy with no zero.

Dynamics is directly associated with the first-ordered dimensions.

Quantum mechanics is directly associated with the second-ordered dimensions.

Equilibrium is not just a process.

Equilibrium is not just a result.

Factors of equilibrium do not imply equality of information.

Factors of non-equilibrium may share much information.

Dynamics has much to do with first-ordered dimensions.

Quantum mechanics has much to do with second-ordered dimensions.

Dynamics, within our gravity drop, is SHFTSS through space and time.

Quantum mechanics is the relationship of space to time.

Chemical clocks are far more accurate than mechanical clocks.

The paucity of explanation in translating Bundal notions of entropy is due to our inability as humans to currently understand entropy from a gravitational point of view.

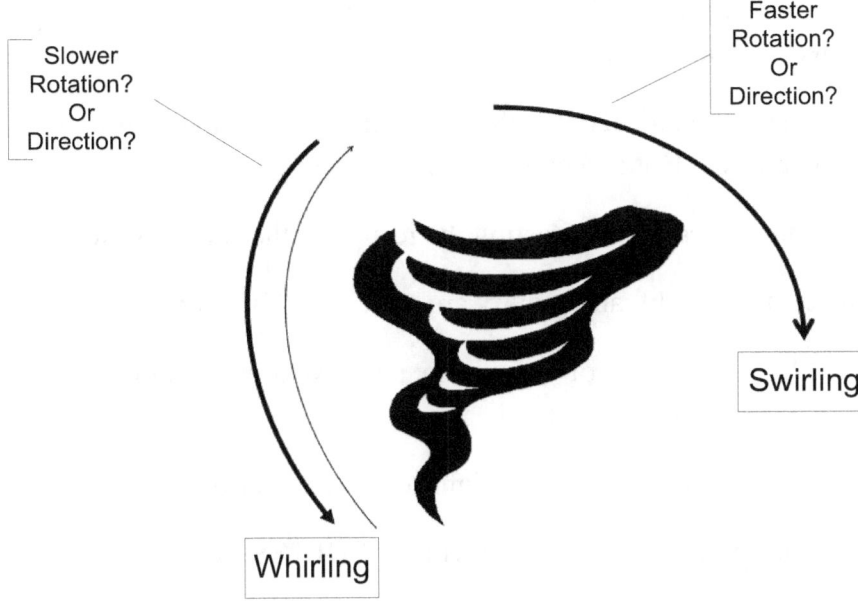

Figure 58: Gravitational whirl and swirl.

Detail: Gravity as a Gerund

The whirl and swirl of gravity gives rise to chirality between the temporal dimensions.

Humans know only one limit to temperature. The classical caloric theory of temperature had the correct insight of a proactive hot and a proactive cold. It failed to understand humans could only see the proactive hot side.

Bifurcation is ontological. Unification is both phenomenological and ontological.

The Bundal model is cosmological.

The funniest law of thermodynamics, imposed to prevent self-references, is the zeroth law.

The second law of thermodynamics is the true nature of all spatial engineers.

Question from the composers: How do the laws of thermodynamics work when the constituency of each dimension has two species separated by other dimensions?

The human ideal of touch, as it applies to thermodynamics needs corrected. Hot and cold are both proactive.

Humans are capable of tasting the results of thermodynamics.

Space, time, light, and gravity are understood by entropy.

The cacophony of sounds in any twentieth-century city expresses entropy as sound.

The cacophony contains significant information.

Entropy is not understood without SHFTSS-ing.

Human Genetics

Reserve for Bundal Log Identification

How can you best identify your best friend?

Endoneurics release and smell endorphins indirectly.

Exoneurics release and smell endorphins directly.

The composers vaguely recall the following contact with Bundals.

The tube was held under our noses. It was not touching our upper lips.

"Now breathe normally. I will not kiss you," the Bundal sighed. "It is difficult to have an initial conversation with someone when you already know what they are like because of their smell."

"What are you talking about," we asked?

The Bundal responded, "Your smell. Your smell gives you away. As you humans say, 'I can smell your genetic makeup. I know who you are.'"

You can smell a book by its cover.

"You have taught us we are not mere chemical factories," we replied. "We make our own choices."

I have been warned about humans. You are going to test me by asking me questions to see how much I truly know about you. The problem is you do not even know yourself, let alone the reality within which you exist. So if I told you about what kind of creature your genetic constituency made you, you would just disagree with certain statements. The irony is I would be correct and it would be you who were mistaken about yourself."

This was as clear and concise as we have ever heard Bundal music being played.

The Bundal was referring to the fact that we humans have an anthropomorphic bias to reality. Some of us are quite convinced the universe was designed specifically for us.

We have learned that genetic makeup is but one part of "who we are." It relates most closely with our spatial material.

There are three other dimensions to reality and each part of "us" relates to these other dimensions. Evolution relates to our temporal process. The level of our health condition relates to our gravitational formation. Our knowledge relates to light. As the four dimensions prism each other our whole emerges.

Unfortunately, as second-ordered dimensional creatures, we are more concerned about our smell and evolutionary status. While our current level of health is important to us, we do not relate it to gravity or phase transitions.

We do relate knowledge and information to light. Twentieth-century humans seem more enamored with silicon than with carbon for information transfer.

Morality
Reserved for Bundal Log Identification

The human idea that genetics is amoral is not completely true. As one part of four dimensions, genetics always has a context. Consider the following as moral versus immoral.

- A fearful creature on the field of battle ...
- A lusting creature on the field of battle ...
- A craving creature on the field of battle ...
- A riled creature on the field of battle ...

Which of these creatures would be considered moral or immoral on the field of battle? You should say it depends on what they actually do.

Whether the creature is fearful, lusting, craving, or riled is immaterial as to what they actually did to contribute to or deny victory for their side. All victors are heroes, the highest level of human morality.

> Now you are admitting the need for the future as well as the past in order to be ethical.

> Humans can only judge based on the past.

The human SHFTSS of the past makes it difficult, but not impossible, for humans to be truly moral as compared to other species. Since we are simply incapable of accurately determining the future, how moral or immoral can we be?

Our laws are based on evaluating the repercussions from past events. But the future changes, and with those changes our laws become invalid if not immoral.

> Would you hang a person for stealing a horse today? When temporally based nuclear tipped bullets can blow up entire cities, will everyone be urged to carry handguns as a deterrent to potential murderers?

> When religion rules governments sometimes it is really good. Sometimes it is really bad. So should religion be involved in governmental policies? The lack of access to the whole of our reality has prevented us from truly appreciating or understanding the proper place of religion for law making.

> Some humans know religion is for love making, not law making.

The Ten Genetic Smells
Reserve for Bundal Log Identification

The following is a revision of the Ten Commandments based upon our genetics. The Bundals were non-committal to our changes. They did appreciate our drive for applying spatial dimension directives to moral issues.

This was one of those rare times the composers did not have to explain things to Bundals. Because Bundals do not count or use numbers as humans do, the issue of why there were ten never came up.

Figure 59: Ten moral precepts based on the past.

Moral Precept	Genetic Smell
Be careful who you chose as a God.	Hunger for direction.
Materialism is great for comfort but not so great to worship.	Hunger for significance.
What you call things is important.	Hunger for information.
Have one day each week that is different.	Hunger for rest.
Treat your parents like you wish they had treated you.	Hunger for wisdom.
Killing is messy in so many ways.	Hunger for survival.
Be careful who you chose as a sex partner.	Hunger for sex.
Figure out socially acceptable ways to eat someone else's food.	Hunger for eating.
When you make things up, at least make other people look good.	Hunger for storytelling.
Trust that you neighbor is going to use their good fortune not to destroy you.	Hunger for shelter.

Bundals can smell the difference when humans start eating to satisfy any of the above ten hungers. Humans can only smell the aftereffects.

How Religionists Respond to Truth

Some humans think there are too many scientific nuts in the world.

For those who believe in science, the questions of truth beguile us. For those who think there are too many religions, read on.

How do I love thee? Let me count the ways.

There are hundreds of disciplines within physics. There are hundreds of sub-disciplines within the disciplines. There are thousands of variations within each subdomain.

Which do you believe?

Their assumptions contradict each other.

Their foundational paradigms are better known as paradoxes.

If you want to hear a prayer just ask a physicist how many dimensions there are.

There are fewer religions than there are types of physical disciplines.

It is amazing how well ancient humans got along with their primitive religious knowledge. Plants, animals, and rocks seemed to have survived and even thrive under human worship …

- … without understanding
- … with understanding
- … with misunderstanding
- … with incomplete and conflicting understanding

The stereo-typical, African witch-doctor can bring more hope and consolation than a myriad of Western, medical specialists who tell

282

people their multiple tests came back normal and yet they are still sick.

Bundals have never figured out the difference between a natural cure and a chemical cure. They thought chemicals were natural by definition.

Humans are always convinced their own doctors are the best in the world.

For some reason some religionists are skeptical of pure, scientific insight.

You do not have to be religious to doubt a scientist.

Bundals want to know: Does toothpaste that passes clinical tests really clean your teeth better than one which does not? Who paid for the tests?

Bundals would just buy the best-tasting toothpaste if they ever needed teeth.

The history of human toothpaste is one of stinging taste. The most popular brand in the early twentieth-century rose to prominence because it stung the mouth. It cleaned no better than anyone else's toothpaste. But its sting meant it had to be good medicine for you. Apparently the stronger the sting the better the clean.

Scientists are sometimes accused of being absent minded. They are not absent-minded. They are simply ignoring the insignificant.

Non-scientists do not realize how much has to be ignored in order to find out what really is going on. Too many facts spoil an experiment. Too many factors make the experiment impossible.

Scientists explain so little yet believe they have explained foundational truths.

Religionists explain too much and believe they have explained foundational truths.

CERN is the largest dice roller ever created by human beings.

You know what scientists say about redoing an experiment: "Just have faith."

Christian people will tell you what they believe and then cement their argument with the phrase, "It is in the Bible." (Of-course it is not.)

Scientific people will tell you what they believe and cement their argument with the phrase, "It's scientific." (Of-course it is not.)

People will pontificate on Einstein and other theories of physics without really understanding the foundations for their pontifications.

What does $E = mc^2$ imply?

Many naively believe scientific specialists understand everything there is to know about electricity.

Educated people know how much they do not know.

The fool believes his or her knowledge is holistic, confusing pragmatism with insight.

Even a scientifically-inclined, male dog knows to lift his leg when peeing. That does not mean he understands the universe.

Twentieth-century humans look at science and religion unfairly.

In science, the unexpected is desired and held sacred. New insights will follow.

In religion, the unexpected becomes mere coincidence often undergirded by false fables involving God. New insights are seen as contrived.

The mere use of numbers convinces us of any truth that fits the scientific fable.

Have you ever noticed how humans have to up-the-ante in order to believe the next-best story? Compare the following stories.

Classical Christian religion has the following stories:

- There is a heaven
- There is a hell
- How you live determines where you live.

Twentieth-century science has the following stories:

- There are 10^{500} numbers of multi-verses.
- Everything is possible and real somewhere.
- Where you live determines whether you live.

Is twentieth-century science trying to up the ante of twentieth-century religion?

Background Noise

One person's noise is another person's clue to the secrets of the universe.

Science is wonderful. It is the assumptions that are in doubt. One great discovery of the twentieth-century is background radiation.

Bundals are again confused with human adjectives since background radiation is all around us and in us.

Nevertheless we can measure the difference between the background radiation over against all of the other radiation in the universe.

Basically we are comparing B-modes to E-modes which are pseudo-scalar fields. Measurements are done by FIRAS on the COBE satellite. Fortunately, this equipment has the ability for real time *in situ* calibration. This mean the results can be adjusted to correct for errors and interference.

Here are NASA's published results for measuring background radiation as found in the Astrophysical Journal, 354: L37-L40, May 10, 1990.

Figure 60: NASA's COBE background radiation results.

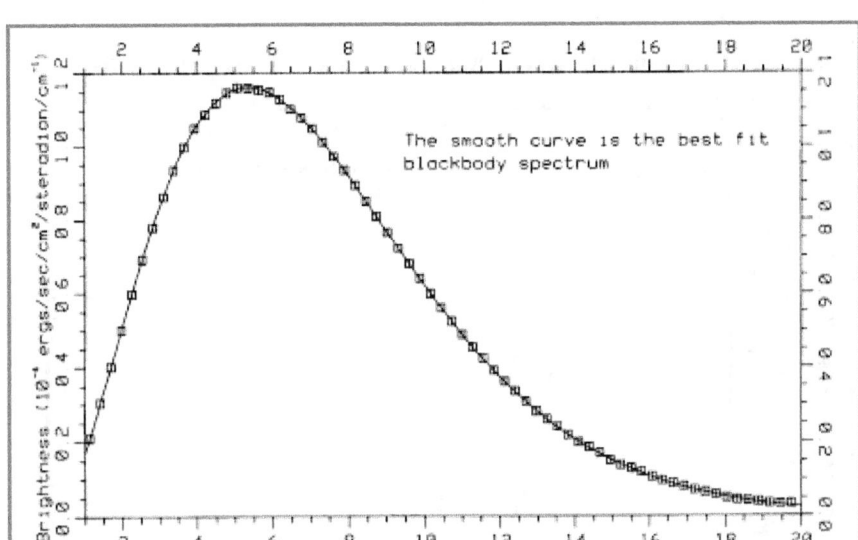

FIG. 2.—Preliminary spectrum of the cosmic microwave background from the FIRAS instrument at the north Galactic pole, compared to a blackbody. Boxes are measured points and show size of assumed 1% error band. The units for the vertical axis are 10^{-4} ergs s^{-1} cm^{-2} sr^{-1} cm.

Other resources, both website and print, comment on NASA's results by reposting the following graph. Notice either graph is almost perfect in comparing expected results with observed results.

Figure 61: A redrawing of COBE background radiation.

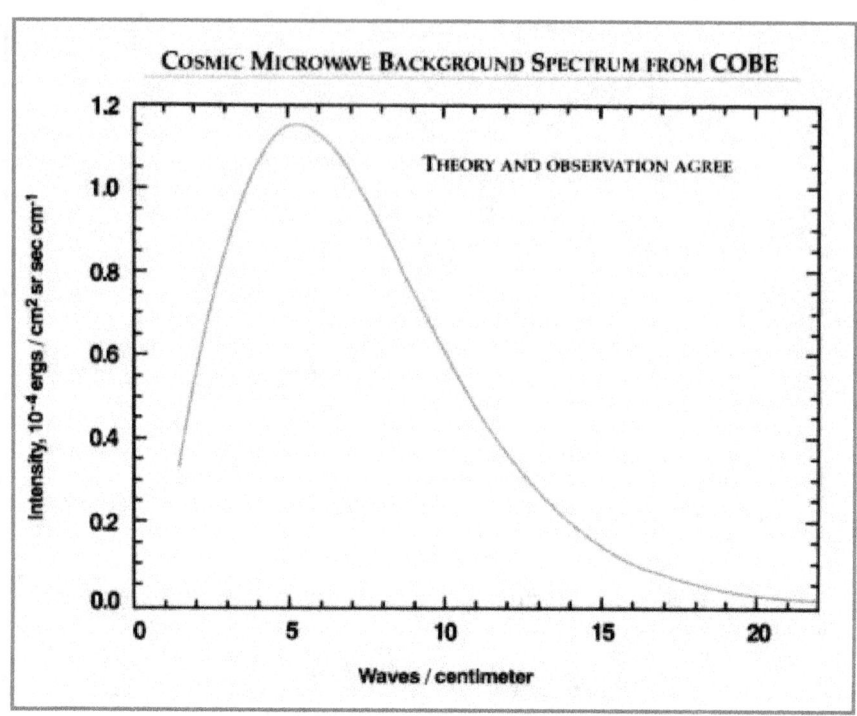

This raises the question as to how thick an Euclidian line has to be in order to hide all errors or degrees of accuracy?

The absence of any apparent error is explained by the fact that the line of theoretical expectation actually hides the actual results obtained from NASA. The line is so thick you cannot even see measurable differences. Those who review the results are astonished and exhilarated. This is obviously the finest fit between observed and theoretical results in astronomical history.

Bundals have questions of the composers.

Are we not suspicious of pseudo-scalars?

Are we not suspicious of instruments that can correct themselves so as to exclude supposed errors in measurement?

Are we not suspicious about any process that uses ratios and then claims to be able to subtract foreground interference from background information?

Are we not suspicious of the assumptions made in the model?

Are we not wondering if we really understand gravity over cosmic differences?

Could FIRAS be refined to discover all the missing Baryons? Perhaps they are hiding behind the background?

One of the composers took chemistry in high school and identifies with the Bundal concerns.

High school textbooks were studied to create equations that would predict experimental outcomes. The class would be turned loosed in a laboratory to actually do the experiment. We would then plot our actual results against the expected results. Our lab results often looked like this.

Figure 62: High school chemists at their best.

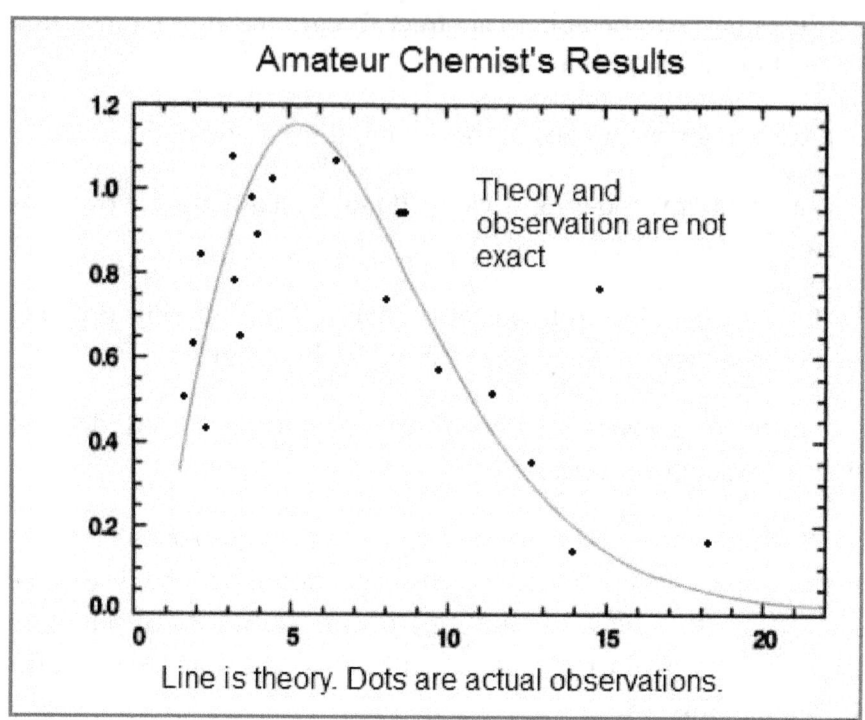

Notice how far off our observed results were with expected results? Most of the class passed with a graph like above.

Two groups of students flunked. They turned in results that exactly matched the expectations. All of their dots were on the line.

No thanks to FIRAS or COBE.

The teacher said, "No true experimental results will ever exactly match expectations. If they do then you have rigged the experiment."

Perhaps there is a difference between chemical experiments in a small laboratory and measuring black-body radiation across the entire universe? Perhaps it is easier to do the latter.

Absolutely reliable results in science surely give humans a reason not to believe in God. Absolute results in religion are seen as propaganda, not truth.

Would the following assumptions change the equations for measuring the temperature of the background noise?

- Matter is created continuously in the universe.
- The idea of a singularity as the source of our universe seems to be a mathematical idealization.
- Is there something behind the background radiation? In other words, is the measurement of the background no more than a measurement of a surface that hides a far more intricate and thicker physical reality?
- What kinds of models are used for measuring and subtracting radiation from interstellar dust, and all of the other parts of the universe not left over?

Humidity

Feelings are forces we misapprehend

A composer's story:

I was in the shower today. It was a dry, mid-Winter day. I turned off the shower. I dried with a towel. I stepped out of the tub and dried my legs and feet. The mirror was foggy. I opened the door a crack to get rid of the fog. I suddenly felt cold.

Was the temperature that much different? No.

> The humidity was diffusing out of the bathroom. The moisture on my skin was evaporating. The evaporation took heat from my skin. I was now colder.

> It felt like cold air. It was really dry air I was feeling.

> I would have said the air was cold. I would have been wrong. The temperature of the air was theoretically the same as what was always in the bathroom. OK. The steam from my shower might have excited some air molecules. Still, it was not so much the temperature change as it was the humidity change the caused me to become cold.

> Feelings can be wrongly interpreted.

The Bundals tell us that temporal energy is the source of three of our forces: The strong force, the electro-magnetic force, and the weak force. Gravity is not caused directly through temporal differentiation. Nor is it caused directly though spatial differentiation. Space-time differentiation is an indirect measurement of constituent elements of the gravitational dimension.

This explains why Newton may need to be revised beyond luminous mass measurements. Luminosity is a direct measure of the gravitational dimension based upon spatial and temporal information flow.

Symmetry-breaking that creates a so-called arrow of time is bosonic. Obviously the Higgs boson and the supposed graviton are intricately related, but not as members of the boson family. Rather they give hint to essential elements within the gravitational dimension.

What feels or measures like the force of a wave of water pushing you back towards the beach, is actually an element of time expressing itself on you. You might say the future has come to you and is pushing you into a new future.

A new type of radical speculation is needed within the sciences and the arts. Human minds are too limited not to explore imagination.

You cannot do away with philosophy.

You cannot do away with history.

You cannot do away with speculation.

You cannot do away with thinking.

It is who we are.

It is what we do.

Even a schizophrenic-paranoid is eerily fascinating. There is some truth to his or her fantasy.

The toughest work we can do as humans is to deconstruct.

It hurts individually and does lead to war.

We will never improve without deconstruction of our cherished ideas and ideals, even if they are the ideas of the whole human race. This is the *Protestant Principle* in philosophical and scientific garb. It may also speak to the deepest aspirations of the religions of the human race.

Temporal energy is the force we all feel.

Of Telescopes and Microscopes

Twentieth-century humans believe they are looking fourteen billion years into their past as they gaze the deepest recesses of space.

The universe was supposedly much smaller then. Why can we not see the smaller universe since we can look so far into the past?

We are told it is because space is expanding. The metaphor is of galaxies living on the surface of a balloon. The balloon is being blown up. The galaxies are still there. They only look farther apart due to the expansion of space.

This explanation admits the past we are viewing is not the past that was.

If space is expanding then our space is expanding.

If our space is expanding then we are bigger today than we were yesterday.

If all of space is expanding then all of space is proportional to what it was before. There should be no distinctive difference among us in proportion to our stars. We should still be in awe of a shrunken, small, initial universe if indeed we are seeing our past in the stars.

It is insisted that gravity holds the galaxies together so apparently these gravity wells have not expanded along with the rest of space. Really? Space has expanded but the stuff in space has not expanded?

Doppler shifts of EM waves from the most distant galaxies indicate they have expanded unless it is *apriori* assumed they have not but only the space between them and us.

295

Tubal's recipe tells us when we gaze at the starts we are seeing our future as temporally projected from our past. We can only see the inside.

When the starts start to fall know for certain good things are not in the future for the planet earth.

As we gaze from our inside we cannot determine extent.

> Our information is limited to light.

> The inside is temporally separated from the outside.

> God has allowed us quite a view of our future. The stars beckon.

The quest to determine what is inside the inside is a canard.

> Fractals tell us there is a finite surface area to the inside.

> Humans are frustrated with the Planck distance. This is the difference between space and time measurements, not just space.

We can only SHFTSS the past-inside with the third M-dimension being a temporal projection into our future.

We can only see a potential-future from the perspective of the inside-past.

> One locked, determined past-inside is followed by another locked, determined past-inside.

> We prehend a three M-dimensional world.

> We apprehend a three M-dimensional world.

> We are suspended in a three M-dimensional world.

Our future-outside enables us to view the past-inside to make changes to the next past-inside.

We do not directly SHFTSS the future-outside.

We imagine the future-outside.

We try to predict the future-outside.

We make differences in the next past-inside.

We make differences in the next future-outside.

We make differences in the next past-inside.

Whether through a telescope or microscope all we are seeing is a potential future.

If a ball is headed toward your head are you seeing your past or your potential future?

If a star, five billion light-years way, is headed towards your head, are you seeing your past or your potential future?

The Headache of Light

God said, "Let there be light," and heavenly springs became earthly strings.

Light is light.

How heavy is a wave?

Generally, relatively light.

Is it not amazing how something that weighs so little pushes us around?

How heavy are particles?

Well, photons have no mass. We have always wondered how something with no mass can strike anything. Would a photon bowling ball knock down any pins?

Particles exist. They leave etchings on photographic plates that prove their existence. You might say particles leave a dent in space, wherever they are.

Photographs of particulate collisions trace the shadows of successive dents. What kind of dent path does a photon follow?

The Einstein-god wrestled with the now. He understood your perspective of simultaneous events is different than my perspective of simultaneous events.

This insight is palpable if humans will accept the existence of now as a fiction.

My future-past is different than your future-past. Agreed? So my perception of now will differ from your perception of now simply because neither now exists. Our differing perceptions are caused by our differing future-pasts.

Ontologically speaking, there is only the past and the future.

The Einstein-god is satisfied to know my past-future is correlated or entangled with your past-future through space, with information carried by light.

Is the information dated?

If it is dated, then why do we need clocks?

If it is not dated, then why do we need clocks?

This correlation or entanglement is not mysterious if one considers the possibility that space and time relate to each other through a third party broker: A dimension of light. This is so silly that it is obvious.

Hyperspace is defined by the limitation of information travel.

Information travel is constructed by gravity.

Gravity shapes hyperspace.

Light shapes gravity.

$E = mc^2$. We have energy and we have mass. Both require a separate third party, light. It is a confusion to relate light as produced within space and time.

Whose idea was it that light was produced within space and time.

Is there no light beyond our light cones?

C. Auguste Dupin found the hidden letter sitting in a card rack, in a public room of the house, safely secured among other letters.

Who would have thought to look there?

Others had torn the house apart, looking for the missing letter.

How many times have humans opened drawers, looking for keys held in their hands? Light is not part of space and time. It informs space and time through gravity.

Ruling out possibilities leaves impossibilities.

Everyone else in the universe, except humans, knows there is no present.

It's obvious.

The future collapses into the next past. Tick. The next past explodes into the next future. Tock.

The Einstein-god scrambled our brains with different clocks moving at different rates depending upon their relative acceleration within the universe.

Or is it just the universe changing its shape?

The Einstein-god could have accepted quantum mechanics if only he realized that the universe has no present. It is just future and past.

Aspirin sellers could create headaches if they just had commercials where physicists explain the difference between mass and weight.

As anyone who has studied special and general relativity knows, it is all about strings.

We meant to say springs. Springs govern clocks and scales, at least until you build your clock with Cesium atoms. Then the springs become rather small. Some say they become strings.

Strings are ontological extensions of the Einstein-god's springs.

Psychologists would say they are a clever, unconscious projection coming from too much study of springs. The technical term is loopy thinking.

Unfortunately, the same problems humans have in trying to understand the Einstein-god's springs occur when you attempt to externalize the springs into strings that create the universe.

This sounds suspiciously like the universe should have two fundamental dimensions: Clock and Scale. How do you use a clock to time itself? How do you use a scale to weigh itself? No wonder we have a measurement problem.

Try timing time. It takes infinite patience. How does something weigh itself, when weight is relative?

"The way to solve this," the Bundals laugh, "is to have ever more numbers of scales or clocks from which to measure." Another way of stating this is to say we increase the number of backgrounds from which we can measure the universe. It is always best to measure with real rulers, even shrinking ones. So let's make each background real. Hence, we can have ten, eleven, sixteen, or thirty-two dimensions, depending on what we are measuring.

If only the Einstein-god had as many clocks and scales.

What a Kaluza.

Sound

Noise cancellation is the mercy of God to humans.

The sounds of soft light reverberate as shrieks of thunder through the gravity fluid. The contracting and exploding of light is SHFTSS throughout the universe.

The gravity fluid crashes in response, whirling and swirling. It implodes and expands with the universe groaning under the pressure. Human mind cannot conceive of the noise.

There is enough noise to form the next new universe. Time and space are born anew. The big bang is always loud.

The reverberations from the past are preserved creating tension within the gravity fluid. The fluid is not allowed to simply evaporate and be gone. It must create anew while honoring the past. The tension is resolved through indetermination. A new future is created with potential for resolution. Decisions will be made and the process of light and gravity will continue with new information to digest.

Time contracts and expands. Space implodes and explodes. The noise is significant even to those outside of space and time.

The drop of gravity fluid quivers. It absorbs information and releases enough to maintain equilibrium as a drop of gravity fluid. It quivers with rest.

Humans are spared in mercy.

> The sounds of the two species of light are cancelled within the inside past.

The sounds of the two species of gravity are cancelled within the inside-past.

The sounds of the two species of time are cancelled within the inside-past.

Only the remaining sound of the inside is heard since it cannot be cancelled by a creature unable to hear the outside.

The sound of the inside is as a whisper to all other sounds.

Humans are wrong when they believe the collisions of atoms and molecules create sound. The sound of a gently blowing breeze through the tree leaves is the residue of creation heard as a comparison of one past moment to another. Such auditory sensations invite a belief in the future and a desire to hear more.

All the sounds of creation cancel each other out leaving only the whisper of the wind. Bose would be proud.

All the flashes of light cancel each other out leaving only the sight we see.

All the forces of creation cancel each other out leaving only what we feel.

All the productions of creation cancel each other out leaving only what we smell and taste.

Taste is the primordial sense. We can navigate all of creation with taste.

All of the human SHFTSS-ing is but a residue left of creation, daring to reveal its true context. Any more of a revelation would terrorize all humans who have ever lived as creatures of space and time. They can only sense the leftovers.

This recharacterizes Plato's shadows on the wall of a cave. Massive amounts of gravity and light engage to cancel each other out. What is left behind is a cave that has humans in it looking at shadows on a wall. Both Plato and Aristotle are vindicated in this model. Both East and West are vindicated in their approach to describing reality. One describes nouns. The other describes verbs.

The aged-old question of whether a falling tree makes a sound in a forest if there is no one there to listen is resolved. The question is based on a human misconception. Any answer will perpetuate the human misconception.

The better question is whether there is a forest without sound.

Tubal's Birth

A new Bundal was born.

It would be many earth-years until Tubal established a sense of Bundal equilibrium. The other intertwined Bundals seem to take longer than usual to regain their equilibrium as well.

Tubal's insistence of an experience with humans was not unique to this newly created Bundals. Many Bundals had strong memories from their intertwining process. The process prepared them for SHFTSS.

Tubal's story concerned the intertwiners.

A long-lived species born in space and time?

A species that could not adequately SHFTSS its own environment?

A species that thought it was the supreme accomplishment of the universe?

A species that killed itself?

A species that doubted the God it knew?

A species that doubted the existence it knew?

This was all too much for the Bundals. Perhaps Tubal needed more time to gain equilibrium.

The Bundals knew new life brings new insights. The Bundals did not procreate for pleasure. New information required new Bundals. New Bundals would serve as increased sources for the stories of old. Information grew.

They knew there is nothing preposterous in life.

They knew there is nothing preposterous in the universe.

The Bundals knew Tubal's experience demanded further investigation. They believed. And in typical Bundal fashion they demanded Tubal keep a log of his journeys on a sea known to Bundals as but a drop of gravity fluid.

And Tubal's log is now your recipe.

Tubal's Recipe

Essential Ingredients:
Reserve for Bundal Log Identification

Start with a sufficient amount of hot and cold light.

Inform a measured amount of wet and dry gravity.

Mix hot and cold light into wet and dry gravity.

Non-Essential Ingredients:
Reserve for Bundal Log Identification

(The following ingredients are non-essential but necessary if the goal is to make the outcome compatible with human taste.)

When the proper consistency of light and gravity is obtained, time and space will be created as the spice of life.

A unique drop of gravity fluid will be formed otherwise known as a universe. It will be filled with two species of space and two species of time.

Consume quickly.

> Tasting this recipe will enable humans to imagine two temporal dimensions.

> Warning: This dietary formula also reduces the number of spatial dimensions to two.

> Enjoy.

Postlude

The humans reveled with their new-found ability to do temporal engineering. Soon they will conquer all of space and time.

The Bundals quiver with laughter.

The human hunger culminates with claiming an entire drop of gravity fluid.

And it is evaporating.